Rudolf Island

Lonely Island

Bear Island

Kilda

Annobón

Ascension Island

Diego García

St. Helena

Tromelin

South Keeling
Islands

Amsterdam Island

Saint Paul Island

Possession Island

Bouvet Island

POCKET ATLAS of REMOTE ISLANDS

THANKS TO
Tim A. Schramm
Borries Schwesinger
Noemi von Alemann
Sandy Weps

JUDITH SCHALANSKY was born in East Germany, near the Baltic Sea in 1980. Judith studied art history at the Free University of Berlin as well as graphic design at the University of Applied Sciences Potsdam. Her typographical compendium on Fraktur, the blackletter font, *Fraktur Mon Amour,* first came out in 2006. It was published by Princeton Architectural Press in the US in 2008, and won several design prizes. Her novel about the naval uniform, *Blau steht dir nicht (Blue is not your colour)*, was published in 2008. When it came to her next book, the *Atlas of Remote Islands (Atlas der abgelegenen Inseln)*, Judith not only wrote the text but also designed and typeset it herself. Since its publication in 2009 it has become an international best seller and won the prize for the most beautiful German book. Judith's latest book, a novel, *Der Hals Der Giraffe (The Neck of the Giraffe)* was published in 2011.

POCKET ATLAS

OF

REMOTE

ISLANDS

BY

Judith Schalansky

Fifty Islands
I have not visited
and never will

Translated from the German
by
Christine Lo

PENGUIN BOOKS

LAND IN SIGHT

Introduction to the paperback edition

LIKE EVERY ATLAS, this one is the result of a journey of discovery. It began three years ago, when I circled the giant globe – as tall as a man – in the Berlin National Library, reading the names of every tiny piece of land marooned in the breadth of the oceans. // They were as full of promise as those white patches beyond the lines indicating the horizon of the known world drawn on old maps. Had our world not already been thoroughly charted, I would have signed on for a voyage of discovery myself, hoping to be the first to see, or even to set foot on, land as yet unknown, and thus to write myself into the atlases of the future. But the days when every ship that sailed round the world uncovered new coastlines and names are long gone. I had to make my discoveries in the library, driven by the desire to find, through rare maps and obscure research papers, my own island: one that I would take possession of, not with the fervour of colonialism, but through yearning for it. // Like everyone else, I imagined my island as an idyllic place, a utopia. My dream promised that if this perfect place could be found, it would be possible to start afresh and do everything differently, far from the pressures of the mainland. It would be a place to find peace, to find oneself, and to finally be able to concentrate on the essentials again.

BUT WHAT I FOUND ON MY JOURNEY were not models of romantic, alternative ways of living, but islands one might wish had remained undiscovered: unsettlingly barren places whose riches lay in the multitude of terrible events that had befallen them. As I found one terrible story after another, I started to drink orange juice by the litre to prevent the onset of scurvy – the disease of vitamin deficiency – that haunted all the reports. At first I was appalled and depressed, but after a while I became agreeably horrified. // As with paintings of the Last Judgement, the eye is arrested by hell, with its terrible monsters and detailed depictions of torture, not by the Garden of Eden. Paradise may be beautiful, but it is not interesting.

QUESTIONING THE TRUTH of what I've written misses the point. There is and can be no clear answer to these questions. I have invented nothing. But I have discovered everything; I have found these stories and made them mine, just as the explorer makes the land he discovers his. All the text in this book has been researched; every detail has been created out of these sources. It is impossible to establish whether events unfolded exactly out of this way, if only because beyond their actual geographical coordinates, islands will always be places we project onto, places which we cannot get a hold on through scientific methods but through literature. // This atlas is therefore primarily a poetic project. Now that it is possible to travel right round the globe, the real challenge lies in staying at home and discovering the world from there.

June 2011, Berlin

*PARADISE IS AN
ISLAND.
SO IS HELL.*

I GREW UP WITH AN ATLAS. And as a child of the atlas, I had never travelled. The fact that a girl in my class had actually been born in Helsinki felt unimaginable. But there it was in her passport: H-e-l-s-i-n-k-i. Those eight letters became a key into another world. To this day, I am baffled by Germans born, for example, in Nairobi or Los Angeles. They might as well claim to come from Atlantis, Thule or El Dorado. Of course I know that Nairobi and Los Angeles exist – they are on the maps. But that someone has actually *been there or even been born there* still feels incredible to me.

I PROBABLY LOVED atlases so much because the lines, the colours and the names replaced the real places that I could not visit. When I was eight, I saw a documentary about the Galapagos Islands. I stared, fascinated by the enormous iguanas with their tiny heads and jagged combs. I still remember the breathless commentary: *Day after day these animals bask motionless on the rocks. This is how the Earth must have looked millions of years ago.* My reaction was immediate: I was going to be a naturalist and travel to these islands. I dragged our atlas down from the shelf and, as the researchers on the television inched their way towards some nesting birds, I heaved open our map of the world. I quickly found the Galapagos. They were a cluster of dots in the light blue ocean. *I want to go there now*, I announced. *Maybe one day*, said my mother, sadly.

But I wasn't to be deflected, and I pushed my index finger across the Atlantic to the tip of South America, turning before the southern polar circle and taking a new direction north at Tierra del Fuego. *Take the Panama Canal, that's shorter*, she recommended, tapping on the line that separated North from South America. And thus I undertook my first voyage round the world.

This map had several colours. The Soviet Union was a bibulous pink. The USA was a reserved blue, nearly as bright as the sea. Then I looked for my country: the German Democratic Republic. East Germans could not travel, only the Olympic team were allowed beyond our borders. It took a frighteningly long time to find. It was as pink and tiny as my smallest fingernail. This was hard to equate: at the Seoul Olympics we had been a force to reckon with, we had won more medals than the United States: how could we suddenly be so infinitesimal?

My love for atlases endured when a year later everything else changed: when it suddenly became possible to travel the world, and the country I was born in disappeared from the map. But by then I had already grown used to travelling through the atlas by finger, whispering foreign names to myself as I conquered distant worlds in my parents' sitting room.

THE FIRST ATLAS in my life was called *Atlas für jedermann* (*Everyman's Atlas*). I didn't realize then that my atlas – like every other – was committed to an ideology. Its ideology was clear from its map of the world, carefully positioned on a double-page spread so that the Federal Republic of Germany and the German Democratic Republic fell on two separate pages. On this map there was no wall dividing the two German countries, no Iron Curtain; instead, there was the blinding white, impassable edge of the page. That, in turn, the provisional nature of the GDR was depicted by the mysterious letters SBZ (*Sowjetische Besatzungszone,* 'Soviet-occupied territory') in the atlases used in West German schools was something that I only found out later, when I had to memorize the rivers and mountains of a home country that had more than doubled in size.

Ever since then, I have not trusted political world maps, in which countries float on the blue ocean like vivid scarves. They grow out of date quickly and give barely any information apart from who is currently running which scrap of colour.

MAPS TELL US MUCH MORE when they do not divide nature into nations; when they allow it to form the basis of comparison across all the borders made by man. In physical topography, land masses glow in the dark green of lowland plains, the reddish brown of mountains or the glacial white of the polar regions, and the seas gleam in every possible shade of blue, quite untouched by the course of history.

In their merciless generalization, these maps tame the wilderness. They reduce geographical variation and replace it with symbols, deciding whether a few trees make a forest or if a human trail is recorded as a path or a track. And thus the width of a motorway is shown to scale, a large city in Germany is depicted with the same square symbol used for one in China, and a bay in the Arctic Ocean shines in the same blue as one in the Pacific because they share the same depth. But the icebergs towering in the Arctic Ocean are ignored.

Geographical maps are abstract and concrete at the same time; for all the objectivity of their measurements, they cannot represent reality, merely one interpretation of it.

THE LINES ON A MAP prove themselves to be artists of transformation: they criss-cross in cool mathematical patterns of meridians and parallel circles regardless of land or water mass, or appear as organic contour lines depicting mountains, valleys and ocean depths. Along with the shading used to create shadow, they ensure that the Earth retains its physicality.

That a finger travelling across a map can be seen as

an erotic gesture became startlingly clear to me when I encountered the three-dimensional counterpart to the atlas in the Berlin National Library: a globe in relief, on which the depths of the Mariana Trench and the heights of the Himalayas became physically graspable for the first time.

The globe is certainly a better representation of the Earth than the collection of maps in an atlas, and it can rouse wanderlust in the young. But the shape of the globe is problematic. Constantly in motion, this Earth has no borders, no up or down, no beginning and no end, and one side is always hidden from view.

IN AN ATLAS, the Earth is as flat as it was before explorers pinned down the white spaces of enticingly undiscovered regions with contours and names, freeing the edges of the world from the sea monsters and other creatures that had long held sway there. In the end, even the giant imagined continent in the southern hemisphere disappeared – but its name had been wrong too: *Terra Australis Incognita*; if a land was unknown, how could it be named?

Depicting a world that could be taken in at a glance was not easy. All the projections provided a skewed picture: either the distances were wrong or the angles or the scale of the surfaces. Thus a representation of the world was arrived at that was true to the angles but shamelessly distorted the size of land masses, so that Africa, the second largest continent, looked the same size as Greenland, the world's largest island, which is actually fourteen times smaller. It is impossible to project the curved surface of the Earth on to a flat surface with the same accuracy for surface areas, distances and angles. The two-dimensional world map strikes a compromise somewhere between impertinently simplifying abstraction and an aesthetic appropriation of the world. In the end, it is simply about grasping the extent of the Earth, orienting it towards the north and being able to gaze down at it like a god. That is how an atlas's supposedly objective view of the world is presented to us, with a scientific appeal to truth that does not shy away from calling a map of the Earth a *world map*, as if no solar system or universe could exist beyond it.

A FEW YEARS AGO, my typography professor showed me an enormous book that she had stored in a huge map chest. I had already seen some of her collection: old poetry albums; watercolours of ribbons and varieties of sausage and cakes; and a long-outdated miscellany with the most tantalizing title: *I'll Tell You Everything*. It was no exaggeration: a display of styles of beard was followed by an overview of human teeth, and the details of ecumenical councils by a list of the most important assassinations of modern times.

But then she brought out a folio of crumpled silk paper wrapped in marbled blue sheets that put even *I'll Tell You Everything* in the shade. Each smooth, yellowed page was full of geometric constructions: crosses, boxes, single, double, triple; broken lines and solid lines; plain, cursive and decorative lettering, abbreviations, arrows and symbols, patches of watercolour and the most delicate cross-hatching. All the protagonists of the cartographical narrative were individually listed and practised in this volume – down to the black and white lines of the borders and the scale measures. Sometimes the stroke of the quill was a little clumsy, but in other places it was so perfect that it seemed barely possible it could have been made by a human hand. The folio was a bound collection of topographical drawings from the apprenticeship of a French cartographer between 1887 and 1889, as the title's ornate majuscule proclaimed.

I discovered a lone, small-format sheet tucked in the endpapers. It was the map of an island in a frame, including the *trompe l'œil* of a sketched fold in the bottom left-hand corner. It had neither a scale nor a description. A brown mountain range rose in a watercolour bulge from this silent, nameless island. In its valleys lay small lakes and rivers snaking their way to a sea that was merely hinted at by the blue contour around the edge of the coast.

I imagined that the cartographer must have had to sketch this island, before he tried his hand at the mainland. It struck me for the first time that islands are in fact small continents, and that continents are in turn no more than very large islands. This clearly outlined piece of land was quite perfect and yet lost at the same time, like the loose sheet of paper on which it had been drawn. Every connection to the mainland had been lost. There was no mention of the rest of the world. I have never seen a lonelier place.

MANY ISLANDS LIE so far from their mother countries that they no longer fit on the maps of that country. They are mostly left out altogether, but sometimes they are granted a place at a cartographical side table, hemmed in by a framed box and with a separate scale measure squashed at the edge, but with no information about where they are. They become footnotes to the mainland, expendable to an extent, but also disproportionately more interesting.

Whether an island such as → Easter Island (174) can be considered remote is simply a matter of perspective. Those who live there, the Rapa Nui, call their homeland Te Pito Te Henua, 'the navel of the world'. Any point on the infinite globe of the Earth can become a centre.

Only when viewed from a continent can such an island – created by active and extinguished volcanoes – be regarded as remote. The fact that it is several weeks' voyage away from the nearest land mass turns it into an ideal place in the heads of continent dwellers, and a land surrounded by water is perceived as the perfect place for utopian experiments and paradise on Earth. In the nineteenth century, seven clans lived in micro-communist harmony under the patriarchal rule of the Scot William Glass on the island of → Tristan da Cunha (74) in the South Atlantic. Dr Ritter, a Berlin dentist tired of civilization and the global economic crisis, set up a retreat on the island of → Floreana (158) in the Galapagos in 1929, where he aimed to renounce all that was superfluous, including clothing. And when the American Robert Dean Frisbie moved to the Pacific atoll of → Pukapuka (150) in the 1920s, he found a remarkable permissiveness – a classic theme of the literature on the South Seas. The island seems to be in its element, still in its natural state, unchanged since the beginning, paradise before the fall from grace, innocent and unblushing.

THE FASCINATION with isolated places also gripped the Californian George Hugh Banning, who sailed the Pacific Ocean at the beginning of the twentieth century, harbouring the desire to be shipwrecked somewhere. He did not care where that might be, *as long as it was a Godforsaken spot surrounded by water.* He became disillusioned when he realized that *we only called at the 'interesting' islands such as Oahu and Tahiti, where chewing gum wrappers and phrases of 'back home' were almost as prevalent as banana husks and the wind in the palm trees.* His luck turned when he joined an expedition in Mexican waters on board one of the first diesel yachts with an electric motor. The voyage took them to the islands of Lower California, including —→ Socorro Island (194), which he knew had barely been visited because there was reported to be nothing there. When asked what was on the island before he set off, he replied, *Nothing. And that is the beauty of it.*

THE ATTRACTION of a beautiful void was also what drew expeditions into the eternal ice (—→ Rudolf Island (40)) to search out the literal nothingness of the poles after the nations who sent their ships out across the world had discovered all the lands rich with vegetation and natural resources and divided them among themselves.

The untouched land of —→ Peter I Island (236) in the Antarctic therefore presented an anomaly, intolerable to our human compulsion to leave tracks across this earth. It also offered the possibility of making a mark on history. Three expeditions failed in their attempts to conquer the island, which is almost completely covered in ice. It was only in 1929 – 108 years after its discovery – that it was landed on, and until the 1990s more people had set foot on the moon than on the island.

MANY REMOTE ISLANDS prove doubly unreachable. The journey to these islands is long and difficult and then landing on them is highly perilous or impossible. Even when it is possible to make landfall, the island that has been the focus of so much yearning often turns out – as might have been expected – to be barren and worthless. Lieutenant Charles Wilkes remarked that —→ Macquarie Island (130) affords no inducement for a visit. Captain James Douglas wrote about the same place: *The island is the most wretched place of involuntary and slavish exile that can possibly be conceived*. Anatole Bouquet de la Grye called the very sight of —→ Campbell Island (166) sad, and George Hugh Banning, the lover of lonely islands, wrote *Socorro seemed a dreary place. The entire island, at the time, put me in mind of a half-burned trash pile, quenched by a shower and left, without even the strength to smoulder, in a puddle of inky rain.*

A LUDICROUS AMOUNT OF EFFORT was often rewarded

with the most meagre results; the majority of the expeditions seemed doomed from the start. The Académie des Sciences sent two expensively equipped expeditions to → Campbell Island (166) at the other end of the world to observe the transit of Venus in 1874, but this natural phenomenon ended up being obscured by a large cloud.

To distract attention from these failures, visiting scientists spend a lot of time measuring every corner of the island or finding specimens of endemic species, listing them in long tables that bulk out the appendices of the expedition reports.

For empirical research, every island is a cause for celebration, a natural laboratory. For once, there is no need to expend great effort to separate the subject under study; the conditions remain recordable and measurable until, of course, the flora and fauna are wiped out by invasive animal species or the indigenous people are struck down by illnesses introduced from foreign shores.

THE FEW VISITORS to these islands are often overcome with horror; faced with the confines of their situation, they are obsessed by the thought that they might be left behind and be forced to eke out their existences on this lonely island for the rest of their days.

The black cliffs of → St Helena (62) became Napoleon's place of exile and death; the fertile green → Norfolk Island (146) turned into the most feared penal colony of the British Empire; and for the shipwrecked slaves of the *Utile*, the tiny island of → Tromelin (108) was at first a lucky escape, but soon their supposed freedom on an island that did not cover even one square kilometre turned into a desperate battle for survival.

A remote island makes a natural prison: surrounded by the monotonous, insurmountable walls of a persistent, ever-present sea, far away from the trade routes which tie overseas colonies to their mother countries like umbilical cords, they are well-suited as places in which to gather everything that is undesirable, displaced and digressive.

In these spaces, terrible illnesses can break out unhindered, and strange customs can hold sway; take, for example, the mysterious infant deaths on → St Kilda (46) and the horrible, yet apparently necessary, practice of infanticide on → Tikopia (206). Crimes such as rape (→ Clipperton Atoll (186)), murder (→ Floreana (158)) and cannibalism (→ Saint Paul Island (84)) seem almost inevitable in these circumstances of such insularity. The fact that even today there are places with laws that go against our sense of justice is shown by the sexual abuse scandal on → Pitcairn Island (178), where the small community of descendants of the *Bounty* mutineers still lives. In

2004, half the adult male population of the island was found guilty of raping the women and children of Pitcairn on a regular basis for decades. In their defence, the accused called upon a centuries-old established right which had allowed their forefathers to engage in sexual relations with under-age Tahitian girls. Paradise may be an island. But it is hell too.

PEACEFUL LIVING is the exception rather than the rule on a small piece of land; it is far more common to find a dictator exercising a rule of terror than an egalitarian utopia. Islands are regarded as natural colonies, just waiting to be conquered. That is how it became possible for a Mexican lighthouse keeper to crown himself king of —> Clipperton Atoll (186) and an Austrian impostor on —> Floreana (158) to proclaim herself the empress of the Galapagos.

Miniature worlds are created on these small continents. Removed from the public eye, human rights abuses can take place (—> Diego Garcia (96)), atom bombs can be deployed (—> Fangataufa (134)) or ecological disasters can be set into motion (—> Easter Island (174)).

There is no untouched garden of Eden lying at the edges of this never-ending globe. Instead, human beings travelling far and wide have turned into the very monsters they chased off the maps.

IT IS, HOWEVER, the most terrible events that have the greatest potential to tell a story, and islands make the perfect setting for them. The absurdity of reality is lost on the large land masses, but here on the islands, it is writ large. An island offers a stage: everything that happens on it is practically forced to turn into a story, into a chamber piece in the middle of nowhere, into the stuff of literature. What is unique about these tales is that fact and fiction can no longer be separated: fact is fictionalized and fiction is turned to fact.

That's why the question whether these stories are 'true' is misleading. All text in the book is based on extensive research and every detail stems from factual sources. I have not invented anything. However, I was the discoverer of the sources, researching them through ancient and rare books, and I have transformed the texts and appropriated them as sailors appropriate the lands they discover.

THOSE WHO DISCOVERED the islands became famous, as if their achievements related to an act of creation, as if they had not merely *found* new worlds but actually *invented* them. Naming geographical features plays an important role in this, as if a place is only brought into existence once it has a name. As at a christening, a bond was created between the discoverer and the discovered, and claiming possession of a land is legitimized, even when the island has only been sighted from afar or has long been inhabited and named.

Scribere necesse est, vivere non est applied, as it does to every action: only that which is written about has really happened. He who sticks the flag into the ground makes sure that he backs up his nation's claim with all manner of information. He calculates the coordinates, makes maps of the land and names the geographical features in his native language. The fact that Norway has created the only up-to-date map of → Peter I Island (236) emphasizes its right to it even though all territorial claims have been suspended under the Antarctic Treaty.

Mapmaking follows on the heels of discovery; and a new place is born with a new name. This foreign land is both occupied and possessed, and the act of conquering it is repeated in the map. Only when a place has been precisely located and measured can it be actual and real. Every map is the result and the exercise of colonial violence.

THE MAP OF AN ISLAND and the land itself occasionally merge, as in the story of August Gissler, for whom the maps he used in digging for treasure for decades on → Cocos Island (214) eventually replaced the gold he searched for. In the end, the promise of the map was worth more to him than the treasure that was impossible to find. It was another hand-drawn map of an island that inspired Robert Louis Stevenson to write his adventure novel: *the shape of it took my fancy beyond expression; it contained harbours that pleased me like sonnets; and with the unconsciousness of the predestined, I ticketed my performance 'Treasure Island'.*

The title of another novel not only became the shorthand for a genre in the vocabulary of literature but also made its way into the atlases. An island in Chile's Juan-Fernández Archipelago was renamed in 1970, purely to attract tourists. On this island, formerly called Más a Tierra ('Closer to Land'), Alexander Selkirk had lived out his *Robinson Crusoe*-inspiring experience. The island does not bear his name today though, but that of his later literary incarnation: → Robinson Crusoe (122). To mix things up thoroughly, another island 160 kilometres to the east, formerly known as Más Afuera ('Further Out'), is now called Isla Alejandro Selkirk, although Selkirk never set foot on it.

Maps shorten the agonizing monotony of a horizon which divides the line of sight on islands day in and day out, against which perhaps the very faint outline of a longed-for ship – a surprising *deus ex machina* – might become visible, bringing food or a return home.

When a newly discovered land does not live up to expectations, revenge can be exacted in the name it is given. Ferdinand Magellan and John Byron named some atolls in the Tuamotu Islands the *Disappointment Islands* in 1521 and 1765 respectively; the former because he found neither the drinking water he sought so desperately nor anything edible there; the latter because the people of the island – now inhabited – were unexpectedly hostile. Many other names are the stuff of myth and fairy tale. The river Styx flows on → Possession Island (92), and the capital of → Tristan da Cunha (74) is called 'Edinburgh of the Seven Seas', though those who live there simply call it 'The Settlement'; after all, how else should you refer to the only settlement within a radius of 2,400 kilometres?

Above all, the names of places reflect the desires and longings of the islands' inhabitants and residents. I have used the latter term in this atlas to demarcate those who live on a particular island temporarily. Those stationed on → Amsterdam Island (100) call one of the capes 'Virgin' and two volcanoes 'Breasts', and a third crater is officially called 'Venus'. Here, the island landscape is ultimately turned into a pin-up poster, an erotic substitute. The island is both a real place and a metaphor for itself at the same time.

IT IS HIGH TIME FOR CARTOGRAPHY to take its place among the arts, and for the atlas to be recognized as literature, for it is more than worthy of its original name: *theatrum orbis terrarum*, the theatre of the world.

Consulting maps can diminish the wanderlust that they awaken, as the act of looking at them can replace the act of travel. But looking at maps is much more than an act of aesthetic replacement. Anyone who opens an atlas wants everything at once, without limits – the whole world. This longing will always be great, far greater than any satisfaction to be had by attaining what is desired. Give me an atlas over a guidebook any day. There is no more poetic book in the world.

ARCTIC
OCEAN

Bear Island

Rudolf Island

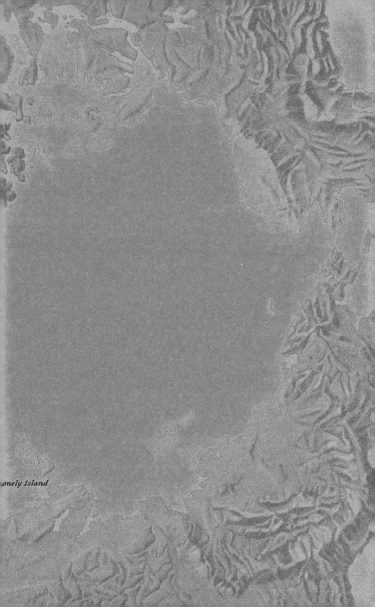

onely Island

^{77° 29' N}
^{82° 30' E} # Lonely Island
(Russia)

NORWEGIAN *Ensomheden*

RUSSIAN *Ostrov Uyedineniya* ['Solitude Island']

20 km² | uninhabited

300 km
--//→ Novaya Zemlya

330 km
--/-/→ Severnaya Zemlya

660 km
--/--/-/→ Rudolf Island (40)

Early 1930s the cervical vertebra of a plesiosaur found on the island

1500   *1600*   *1700*   *1800*   *1900*   *2000*
-/-

26. Aug. 1878 discovered by Edvard Holm Johannesen

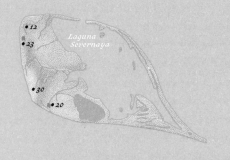

Laguna
Severnaya

12
23
30
20

0   1   2   3   4   5 km
/--/--/--/--/--/--/

Lonely Island

LONELINESS lies in the centre of the Kara Sea in the northern Arctic Ocean. This island is worthy of its name: it is cold and barren, trapped in pack ice all winter, with an average annual temperature of 16 degrees; at the height of summer the temperature sometimes rises to just over freezing. // No one lives here. A former polar observatory has sunk into the snow and abandoned buildings doze in the belly of the bay, facing the narrow spit of land beyond the frozen marsh. // A prehistoric dragon's skeleton was found here. Some years later, a German submarine fired grenades at the weather station, destroying the barracks and killing the garrison. Firing at Lonely Island was one of the final manoeuvres of the German navy's 'Operation Wunderland' in the Second World War. // The observatory – one of the Soviet Union's largest – was rebuilt during the Cold War. The name that the Norwegian captain from Tromsø had given the island was forgotten – in Russian, Lonely Island became Solitude Island. The visitor to the island was now not a prisoner but a hermit, sitting out the years in an icy wasteland until he could return to the mainland a saint. // The provisions that have been left behind are in deep freeze in the green wooden barracks, all crusted with ice, as are the instruments once used to measure air pressure, temperature, the direction of the wind, the height of the clouds and the radiation from the skies. The rain gauge lies buried in the snow. A photograph of a goateed Lenin hangs against palm-patterned wallpaper. The logbook contains precise records of the chief mechanic's maintenance work: the levels of oil and petrol in every machine. The final entry in red felt-tip spills over the confines of the columns: *23 November 1996: The evacuation order came today. Pouring the water out. Turned off diesel generator. The station is* ... The final word is illegible. Welcome to Lonely Island.

Bear Island

Spitzbergen (Norway)

NORWEGIAN *Bjørnøya*

178 km² | 9 residents

220 km
--/—→ Spitzbergen

390 km
--/-/—→ Norway

2160 km
--/--/--/--/-/--/--/--/--/-/—→ St Kilda (46)
1000

1898 reclaimed by Theodor Lerner for the German Reich
1500   1600   1700   1800   1900   2000
-/--/
1920 annexed by Norway
10.Jun.1596 discovered by Willem Barents and Jacob van Heemskerk

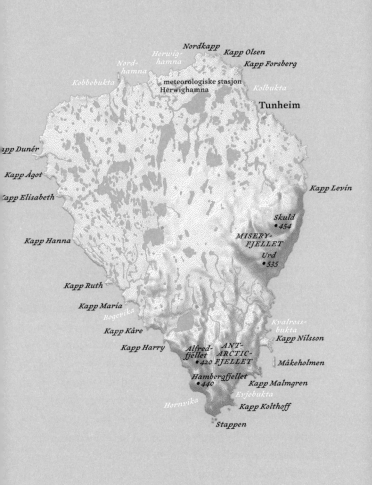

Nordkapp

Herwig- · Kapp Olsen
hamna
Nordhamna · Kapp Forsberg

Kobbebukta
meteorologiske stasjon · Kolbukta
Herwighamna

Tunheim

Kapp Dunér

Kapp Ågot
· Kapp Levin

Kapp Elisabeth

Skuld
· 454
MISERY-
Kapp Hanna · FJELLET
Urd
· 535

Kapp Ruth

Kapp Maria · Kvalross-
Bogevika · bukta
Kapp Kåre · Kapp Nilsson
Kapp Harry · Alfred- · ANT-
fjellet · ARCTIC- · Måkeholmen
· 420 · FJELLET
Hambergfjellet · Kapp Malmgren
· 440 · Eryebukta
Hornvika · Kapp Kolthoff

· Stappen

0 1 2 3 4 5 km
|—|—|—|—|—|—|—|—|—|—|

Bear Island

IT IS OVERCAST but the barometer reading is high. They arrive at the south harbour of Bear Island at two in the morning on 30 June 1908: seven obsessive bird-watchers on the steamship *Strauss*, along with four taxidermists and a gunsmith. Baron Hans von Berlepsch, the founder of the bird protection movement, is standing on deck with a pair of binoculars slung round his neck and, by the grace of Barbarossa, five parakeets on his coat of arms. Silent, he listens in the dark to the song of the birds that he has only known from books until now. // By the morning the gentlemen have shot fulmars and guillemots, a young ivory gull and a fully grown great black-backed gull. Swarms of newly hatched glaucous gulls are walking up and down the beach. The bird enthusiasts grab a handful of chicks, still covered in grey down, and take them on board: two to raise themselves, the others to be killed and skinned. The auks watch from the breeding cliffs. // Someone bags a lesser black-backed gull which, on closer examination, turns out to be a small herring gull. Someone else outwits a red-throated diver. Inland, they track down a long-tailed skua, and even find some black scoter on lake ice. On the gravel bed of a small stream they shoot a female ringed plover, and a pair of snow buntings flap around them in such alarm that they betray the location of their nest – sadly still empty. A pair of arctic skuas try to distract attention from their nesting place through a display of flight. But they find the eggs in a mossy hollow, camouflaged olive, with dark speckles. The bird baron collects four whole clutches of eggs and one half-clutch, and carries them on board in handkerchiefs. The other gentlemen catch sight of the coveted razorbill among thousands of guillemots. Shots ring through the air and a specimen crashes dead to the ground in all its feathered glory. The evidence has been produced; its existence on the island has been proven. The bird enthusiasts are content. While they size up their prize, a flock of gulls devours the remains of a whale carcass on the beach.

81° 46' N
58° 56' E

Rudolf Island

Franz Josef Land (Russia)

Also known as *Crown Prince Rudolf Land*

RUSSIAN *Ostrov Rudolfa*

297 km² | uninhabited

560 km
--/--//⟶ Severnaya Zemlya

590 km
--/--/-/⟶ Spitzbergen

1170 km
--/--/--/--/-/⟶ Bear Island (36)

1500   1600   1700   1800   1900   2000
-/--/--/--/--/--/--/--/--/--/--/--/--/--/--/--/--/--/-/--/--/--/-

Apr. 1874 discovered by Julius Payer and Carl Weyprecht
on the Austro-Hungarian North Pole expedition

Rudolf Island

THE SLEDS are travelling north in -50 degrees, laden with thirty pounds of bear meat, heading towards the next latitude. The bleeding paws of the sled dogs stain the snow. Glaciers glisten and crack in the sunshine. The landscape is barren, bare and white, like the map. The atlas has few blank spaces now; the last are waiting to be labelled at the edge of the world: the no-man's land with no cardinal points. The silent place which determines the direction of the compass needle has not been reached, so the riddle of the Northwest Passage remains unsolved: the dream of an open sea warmed by the Gulf Stream, a navigable passage, an ice-bound route to India. // Leaving the sleds behind, they sleep in glacier fissures and continue northwards on foot, led by Lieutenant Julius Payer, who was the first man to climb over thirty Alpine peaks and is now the commander of the expedition in this country. But this is not a country. It is another island that he has named like a country, just as he has done with the whole of this newly discovered archipelago. Payer is never at a loss when it comes to names; he christens islands, glaciers and rocky projections tirelessly, after the home towns of the sweethearts of his youth, after his patrons, his colleagues, archdukes and the Austrian Empress Sissi's son. He carries his homeland out into the ice: using the names of his country's fathers, in the name of the fatherland. // The compass shows that they have crossed the 82nd parallel north, a further invisible line in the snow which the lieutenant records on his silent map. In the evening, they reach the edge of Crown Prince Land. What lies before them is not a navigable sea, but a gigantic open expanse surrounded by old ice. Mountainous clouds shimmer on the horizon. The lieutenant sketches flowing lines on the piece of paper one last time: Cape Felder, Cape Sherard Osborn and the southern tip of Petermann Land. They drive the Austro-Hungarian flag into the rocky ground and cast a bottle containing a message off a cliff ledge. Words frozen for future witnesses: *Cape Fligely, 12 April 1874, 82°, 5°, northernmost point. Thus far and no further.*

ATLANTIC
OCEAN

Brava

Trindade

Tristan da Cunl

Southern T

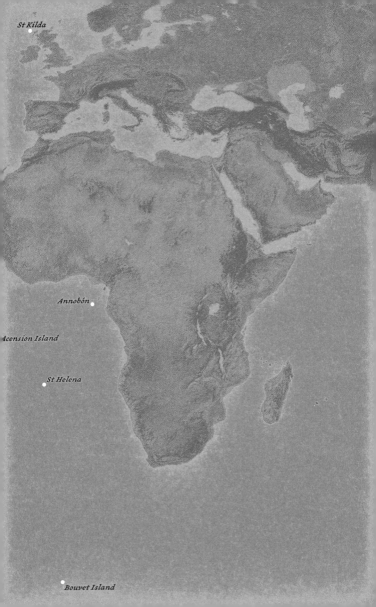

57° 49' N
8° 35' W
St Kilda
(United Kingdom)

GAELIC *Hiort* or *Hirta*

8.5 km² | uninhabited

60 km
/→ Isle of Harris, Outer Hebrides

160 km
-/→ mainland Scotland

1000   *2000*   *3000*   *4940 km*
--/--/--/--/--/--/--/--/--/--/--/--/--/--/--/--/···/→ Brava (54)

1850s wave of emigration to Australia

1500   *1600*   *1700*   *1800*   *1900*   *2000*
-/-

1826/27 smallpox epidemic   *1930* evacuation

1891 last recorded case of neonatal tetanus

Stac an Armin

• Mullach an Eilein
379
Stac Lee •
BORERAY

Am Plasdair
Glen
Bay
SOAY
Conochair
430 •
Mullach
Bì • 355
Village
Bay
HIRTA
DÙN
Gob an
Dùin
Stac Levenish

0 1 2 3 4 5 km
|--|--|--|--|--|--|

St Kilda

ST KILDA – YOU DON'T EXIST. Your name is just a faint cry made by the birds that make their home on the high cliffs at the furthest edge of the United Kingdom, beyond the outermost of the Outer Hebrides. Only when a north-east wind prevails can the voyage even be attempted. // There are sixteen cottages, three houses and one church in the only village on St Kilda. The island's future is written in its graveyard. Its children are all born in good health, but most stop feeding during their fourth, fifth or sixth night. On the seventh day, their palates tighten and their throats constrict, so it becomes impossible to get them to swallow anything. Their muscles twitch and their jaws hang loose. Their eyes grow staring and they yawn a great deal; their open mouths stretch in mocking grimaces. Between the seventh and the ninth day, two-thirds of the newborn babies die, boys outnumbering girls. Some die sooner, some later: one dies on the fourth day, another not till the twenty-first. // Some say it is the diet: the fatty meat of the fulmars and their eggs smelling of musk that make the skin silky smooth but the mothers' milk bitter. Or that it is a result of inbreeding. Yet others say that the babies are suffocated by the smoke from the peat fires in the middle of the rooms, or that it is the zinc in the roofs or the pale pink oil that burns in the lamps. The islanders whisper that it is the will of the Almighty. But these are the words of pious men. The women who endure so many pregnancies and bear so few children who survive the eight-day sickness remain silent. // On 22 June 1876, one woman stands on the deck of a ship that is bringing her home. Like all the women of St Kilda, she has soft skin, red cheeks, exceptionally clear eyes and teeth like young ivory. She has just given birth to a child, but not at home. The wind is blowing from the north-east. Long before she can be seen from the shore, she lifts her newborn high in the air.

Ascension Island

(United Kingdom)

PORTUGUESE *Ascensão*

91 km² | 1,100 residents

1000   *1560 km*
--/--/--/--/--/--// → Ivory Coast

1000   2000 *2250 km*
--/--/--/--/--/--/--/--/--// → Brazil

1000   *i*   *2110 km*
--/--/--/--/--/--/--/--/-/ → Trindade (66)

20.*May 1503* (*Ascension Day*) rediscovered by Afonso de Albuquerque

1960–61 missile-tracking station is built

1500   1600   1700   1800   1900   2000
-/-/

25.*Mar.1501* discovered by João da Nova

15.*Dec.1899* the first undersea cable is laid

North Point

English Bay

Pyramid
Point

Comfortless
Cove

Broken
Tooth
• 226

Porpoise
Point

North East Bay

Hummock Point

Clarence Bay

Georgetown

Catherine
Point

Sisters
Peak
445 •

Cross
Hill

Two Boats
Village

Boatswain Bird
Island

Lady Hill
• 329

• The Peak
• 859

White Hill
• 525

Main Base

GREEN MOUNTAIN

Whale Point

McArthur Point

Round Hill

• Dark
Slope Crater

South
East
Bay

Unicorn Point

Portland
Point

WIDEAWA
FAIR

Crystal Bay

Cocoanut
Bay

Mars Bay

Pillar
Bay

Gannet Bay

South Point

0 1 2 3 4 5 km
|--|--|--|--|--|--|

Ascension Island

EVERYTHING is reaching for the sky: the forty-four sleeping craters in the rust-red cone of ash, the antennae several metres long and the spreading satellite dishes. They are eavesdropping on the continents, listening to the world, to the universe, to the infinity of outer space. This is a wasteland of cooled lava, inhospitable as the moon. The whitewashed church of St Mary sits at the foot of dusty Cross Hill like God's final fortress on Judgment Day. // No one lives on Ascension Island permanently – everyone is there to work, and no one will be allowed to stay. This desolate country is unsaleable. It is a working island for telecommunications operators and spies, and a landing site for the cables that run along the floor of the Atlantic Ocean to connect the continents. NASA stretches out its feelers here, building a tracking facility for intercontinental missiles and scattering glistening white parabolic antennae all over the land, oversized golf balls stuck to the edges of the craters. // On 22 January 1960, the *Atlas* is launched into space from Florida but re-enters the Earth's atmosphere just before Ascension Island. Richard Aria, a technician with Cable and Wireless, is watching the sky over Red Hill. Nothing. Only the Big Dipper, which is upside down here. Half an hour passes – still nothing. Suddenly there are two green flashes – there it is! The missile races towards the Earth in an iridescent streak, lighting up the whole island: first green, then yellow, red, orange, then green again – heading ever downwards until it is extinguished. Parts rain down from the end, glowing fiery red, and its nose glitters as it burns up in the sea in shifting colours: a startling red, crimson, a dull dark red. Then nothing but blackness until a long, deep drone rises from the sea, followed by a deafening explosion and a thundering boom that lasts for at least a minute and a half. Until an American voice trumpets, *Get a load of that, you — Russians!* The space race has begun on Ascension Island.

14° 51′ N
24° 42′ W

Brava *Islands Under the Wind* (Cape Verde Islands)

PORTUGUESE *Brava* ['Untameable']

64 km² | 6,804 inhabitants

20 km
/→ Fogo

780 km
--/--/--//→ Dakar

1000   2000   2760 km
--/--/--/--/--/--/--/--/--/--/--/--/--//→ Ascension Island (50)

1460s discovered by Portuguese sailors

/ 1500   1600   1700   1800   1900   2000
-/-

1573 first settlement

1680 volcano erupts on neighbouring island Fogo

Ilhéu de Cima

Ilhéu Grande

ILHÉUS DO ROMBO

Ponta do Incenso

Ponta da Vaca

Porto do Sorno

Ponta Jalunga

Ilhéu da Areia

Furna

Vila Nova Sintra

Santa Bárbara

Fajã d'Água

Mato Grande

N. S. do Monte

Ponta Mínhoto

Ponta da Costa

Fontainhas

•976

Baía do Caniço

Campo Baixo

Mamama

Ponta de Rei Fernando

•831 Cachaço

Ponta do Tambouro

•673

Ponta de Morea

Monte da Ponta Verde

Porto de Tantum

Ponta Façanha

Ponta do Alto

Cova de Mar

Porto de Ançião

Ponta Nhô Martinho

0 1 2 3 4 5 km
|---|---|---|---|---|

Brava

THIS CLENCHED HEART lies untameable, protected from the wind by the great volcano of the neighbouring island. Here, at the outermost edge of the archipelago, the clouds hang low and it rains more than on the other islands, which are continually battered by desert winds. Dew forms on the leaves of the almond trees and the date and coconut palms, on the petals of the fringing lobelia, oleander, hibiscus, jasmine and bougainvillea. This island has veins of rivers and strong muscles in its mountain range. The faint beat of the melancholy *morna* sounds, and the old song pulses relentlessly in a minor key, a lament about the inescapability of fate. It is the longing for an unnameable moment in the past, for a distant land, for a long-lost home. A feeling, scattered like these islands, the yearning for a place that is at once everywhere and nowhere. This is the song of a land without original inhabitants. Everyone who lives here is descended from the planters who stayed behind and from their slaves, from those who chose to move here and those who were forced to, a people with blue eyes and black skin. // The melody starts hesitantly, following the wide arc of a legato. The guitar adds a bass line in four-four time, accompanied by the plucking syncopations of the cavaquinho, sometimes backed up by a violin. These songs live in the bars and dance halls of the harbour: *Who goes with you / on this long journey? / Who goes with you / on this long journey? // This journey / to São Tomé // Homesick, homesick / homesick / For my country São Nicolau // When you write to me / I will write to you / When you forget me / I will forget you // Homesick, homesick / homesick/ For my country São Nicolau // Until the day / that you return.* // Two-thirds of Cape Verdeans do not live on the islands any more.

^{1° 26' S}
^{5° 38' E} # Annobón

(Equatorial Guinea)

PORTUGUESE *Ano Bom* ['Good Year']

FA D'AMBU *Pagalu* ['Big Cockerel; Parrot']

17 km² | 5,008 inhabitants

190 km
--/⟶ São Tomé

610 km
--/--/-/⟶ Malabo

     1000   *2000*   *3000*   *5580 km*
--/--/--/--/--/--/--/--/--/--/--/--/--/--/--/--/···/⟶ Tromelin (108)

1470 discovered by Diego Ramirez de la Diaz
/ *1500*   *1600*   *1700*   *1800*   *1900*   *2000*
-/--/--/--/--/--/--/--/--/--/--/--/--/--/--/--/--/--/--/-
                                        /
1968 became part of Equatorial Guinea

San Antonio
de Palé

Isla Tortuga

Dyo Dyo

Pico do Fogo
• 435

Pico
Quioveo   Anganchi
• 598

Bahía
de Aual   Bahía de
                            A Jabal
Aual

Mábana   Punta
                Olonganchi

Punta Manjob   A Dyibó

0   1   2   3   4   5 km
|---|---|---|---|---|

Annobón

ON 26 SEPTEMBER 2003, 3COV goes on air. Though the weather is poor and they have not even activated the lowest frequency band, they already have a great many radio contacts. The lower the frequency, the longer the wavelengths. // The army bothers them every day, asking questions and wanting to see their papers, even though they wrote to the country's authorities in advance to make it clear that amateur radio operators are interested neither in political nor in religious matters, only in cross-border communication. Every parcticipant has personally obtained permission from the Minister of Transport and Communication to visit the island for two weeks, and has a special dispensation from Customs to enter and exit with radio equipment. // At ten o'clock on 4 October, the expedition ends abruptly. Official sources have ordered that all transmissions stop immediately and all antennae be retracted. The hams are given three hours to dismantle their station, and are flown to the capital city of Malabo in a Russian cargo plane the same day. They are unable to save much of their photographic material, and their telephone conversations are repeatedly interrupted. // DJ9ZB and EA5FO are allowed to leave two days later. EA5BYP and EA5YN are detained. On 10 October, they, too, are finally allowed to leave for home. *We deeply regret that we have been unable to achieve the objectives of this expedition, and are very grateful for the help given us by associations, clubs and individuals, as well as the kindness and friendship always given us by the inhabitants of Annobón. We do not feel free to share any further details about what happened, in order to keep open the possibility of future expeditions to Annobón. We hope you understand the difficult and delicate situation we went through. Nevertheless, we do not abandon hope of reactivating 3COV when circumstances improve.* Roger, over and out.

St Helena

15° 57' S
5° 42' W

(United Kingdom)

122 km² | 4,255 inhabitants

1000   1850 km
--/--/--/--/--/--/--/--/--/--/→ Angola

1000   2000   3000 *3290 km*
--/--/--/--/--/--/--/--/--/--/--/--/--/--/--/--/--//→ Brazil

1000   2010 km
--/--/--/--/--/--/--/--/--/--//→ Annobón (58)

5. May 1821 Napoleon dies

1500   1600   1700   1800   1900   2000
-/-

21. May 1502 discovered by João da Nova

15. Oct. 1815 Napoleon arrives

Sugar Loaf
Point

Buttermilk
Point

*FLAGSTAFF
BAY*

Flagstaff 818 Barn Long Point

Jamestown *DEADWOOD* •688 The Barn

PLAIN *TURK'S CAP*

Half Tree *VALLEY*

Hollow The Black Point

Lemon Valley Bay Briars 406 • *Prosperous*

DONKEY Horse Point *Bay*

Long Ledge *PLAIN* Longwood

Glencot

Horse Pasture *HORSE* *PROSPEROUS* *Dry Gut*

Point *PASTURE* *BAY PLAIN* *Bay*

Egg Island Mount Actæon Great Gill Point

• 820 Stone Top *Stone Top*

Thompson's •707 •798 494 • *Bay* George Island

Bay High Hill High Peak White Hill Long Range

outh 543 • • 588 *Deep Valley*

t Point White Point *HORSE* *Bay*

694 • *RIDGE*

Old Joan • 691 Powell

Point Lot's Wife *Bay*

MANATI • 462 *SANDY*

BAY Great Hollow *BAY*

• 573 White Bird Island

Speery Island Robert Rock

0 1 2 3 4 5 km
|---|---|---|---|---|

St Helena

ONE FRIGATE *is decidedly not enough!* exclaim the Bonapartists, and demand a whole fleet. Ultimately, this is about reparation for what had been lost at Waterloo. But it is not the old ferryman Charon but the young Prince de Joinville who is commanding this mission. The funeral procession consists of a royal commissioner, a priest, a doctor, a locksmith, a graphic artist; the official escort provided by a few faithful followers and servants from his exile. All of them travel across the sea to collect the body of the man whom Europe had wanted to keep far away. The frigate *Belle Poule* is painted black especially for the voyage to the island of death. // Napoleon had always failed with islands. Not one battle at sea had he won. Perfidious Albion! It was not liberty that he lacked on this island but power, the prospect of a return to the world stage. Guarded by a regiment, he resided on a high, windy plateau with a circle of his loyal traitors. All he could do was become a martyr, and his followers wrote gospels for him, while he played Prometheus against the blackened rocks and listened to the echoes of his past, which had curdled into history. // At midnight exactly, British soldiers wrench the iron grate and the three slabs free of the earth. By the light of a torch, they uncover the four nesting coffins of mahogany, lead, ebony and tin. They open the final one carefully, and the doctor lifts the white linen cloth away. There he lies in the uniform of the Chasseurs de la Garde, medals on his breast, hat resting on his thighs as if asleep: quiet and relaxed, with a deformed nose, a bluish beard and long, very white fingernails. His body is dessicated. Those who disturb the sleep of the dead are shocked; his loyal followers weep. // In the pouring rain, a procession of forty-three carries his sarcophagus to the road, where it is heaved on to a wagon and covered with a purple pall embroidered with golden bees and majuscule Ns. // Three days later, on 18 October 1840, anchor is weighed. The emperor is returning home.

Trindade *Trindade and Martim Vaz* (Brazil)

PORTUGUESE *Ilha da Trindade* ['Trinity Island']

10 km² | 32 residents

```
                    1140 km
--/--/--/--/-/ —→ Vitória

                 1000    1450 km
--/--/--/--/--/--/ —→ Rio de Janeiro

           1000              2000    2540 km
--/--/--/--/--/--/--/--/--/ —→ St Helena (62)
```

```
                                    1890–96 occupied by Great Britain
     1500      1600      1700      1800  /  1900        2000
-/--/--/--/--/--/--/--/--/--/--/--/--/--/--/--/--/--/--/--/-
 /
18. May 1502  discovered by Vasco da Gama
```

Ponta da Norte

Ponta Crista de Galo

Obelisco
● 430

Ponta do Monumento

Enseada
dos Portuguêses

Pico Desejado
620 ●

Pico Branco
● 470

Parcel
das Tartarugas

Enseada
da Cachoeira

Ponta dos Cinco Farilhões

Enseada
do Príncipe

0   1   2   3   4   5 km
|---|---|---|---|---|

Trindade

THIS PLACE is a topographical disaster. Everything has been arbitrarily hurled into the ocean; the ground is rutted, downward sloping and hostile. Over and over again, someone out on a walk disappears without trace – washed away by waves several metres high, buried alive by a landslide or swallowed up by a crater. In the cemetery, crosses without graves stand memorial to those who have disappeared. // At midday on 6 January 1958, shortly before the research vessel *Almirante Saldanha* weighs anchor, Almiro Barauna, one of the civilians on board, is taking a few more photographs of the south coast of Trindade, when, at fifteen minutes past twelve, a brightly lit object appears in the sky, moving towards the island, heading for Ponta Crista de Galo in bat-like, wave-motion flight. // The flying disc has a metallic glint and is surrounded by a greenish phosphorescent mist. In uproar, the officers and sailors on deck point at the glittering object. Thirty seconds pass before Barauna finally takes up his camera, looks through the viewfinder and presses the button twice; then the object dips behind the peak of Pico Desejado. A couple of seconds later, the flying object, which has clearly flown in a loop, is visible again. It seems nearer than before, and larger. All is confusion on the navigation bridge. Barauna stumbles in the mêlée, but takes four more photographs before, approximately ten seconds later, the mysterious object disappears into a distant bank of clouds, for ever this time. // Barauna's photographs are overexposed. Four of the six photos show the unknown object in different flying positions. With a ring round its middle, it looks like Saturn pressed flat. Two of his shots have fallen victim to the tumult on board: they show nothing but a slant of railing, water and the dark rock of a coast that rises from the sea in rigid points, alien and sinister, as if from another world.

^{54° 25' S}
^{3° 21' E} # Bouvet Island

(Norway)

NORWEGIAN *Bouvetøya*

ENGLISH formerly *Lindsay* or *Liverpool Island*

49 km² | uninhabited

```
              1000        1700 km
--/--/--/--/--/--/--/--/→ Antarctica

              1000        2000    2510 km
--/--/--/--/--/--/--/--/--/--/--/→ Cape of Good Hope

              1000        1910 km
--/--/--/--/--/--/--/--/-/→ Tristan da Cunha (74)
```

```
   1.Jan.1739 sighted by Jean-Baptiste Charles Bouvet de Lozier
                              /        27.Feb.1930 annexed by Norway
   1500      1600      1700  /    1800      1900  /    2000
-/--/--/--/--/--/--/--/--/--/--/--/--/--/--/--/--/--/--/-
              10.Dec.1825 landed on by George Norris
```

Kapp Valdivia

Kapp
Circoncision

VICTORIATERRASSE

Olavtoppen
•780

Kapp Lollo

WILHELMPLATÅET

SLAKHALLET

Kapp Meteor

ESMARCHKYSTEN

•766
Lykke
toppen

Kapp Norvegia

VOGTKYSTEN

Kapp Fie

Larsøya   Catoodden

0   1   2   3   4   5 km
/---/---/---/---/---/

Bouvet Island

A WIDE EXPANSE of sea stretches south of the Cape Province, still unexplored by oceanographers. Right after the Agulhas Bank, all soundings break off. Painted a tropical white, the *Valdivia* steers its way southwards, taking a course that no ship has chosen in over fifty years. The area is not described on British sea charts; there is just one uncertain entry: a small archipelago beneath the 54th parallel, sighted by Bouvet, who took it for the cape of a continent to the south. Neither Cook nor Ross, nor Moore, found it again. Two whaling boat captains claim to have seen the island, but each give different co-ordinates. // The barometer drops and the wind rises to a storm, ten on the Beaufort scale, forcing them to heave-to. The skies darken and the petrels take position; the first ash-grey albatrosses with blackened heads and white-rimmed eyelids circle above the struggling ship in silent, ghostly swoops, like vampires. Caught by the swell, the steamship is tipped on its side many times, so sharply that the glass flasks in the laboratories crash out of their holders. The steam whistle sounds over and over again, and the icebergs hiding in the mist reply with their high, clear echoes. At last, the *Valdivia* arrives at the area shown as three islands on the Admiralty chart: Bouvet, Lindsay, Liverpool. Soundings indicate an undersea ridge, and the sun shining on the wall of clouds on the horizon creates the illusion of land. There is no sign of the island. // At midday on 25 November 1898, the first large iceberg hoves into view, gleaming majestically. At thirty minutes past three, the first officer cries, *The Bouvets are in front of us!* But what appears first in hazy, then in increasingly clear, outline only seven nautical miles to starboard is not a group but a single steep island in all its wild glory, with sheer walls of ice and glaciers cascading to sea level; an enormous field of firn. This is it, Bouvet Island, sought in vain by three expeditions, missing for nearly seventy-five years.

Tristan da Cunha

37° 6' S
12° 17' W

(United Kingdom)

104 km² | 264 inhabitants

1000 2000 **2770 km**
--/--/--/--/--/--/--/--/--/--/--//→ Cape of Good Hope

1000 2000 3000 **3340 km**
--/--/--/--/--/--/--/--/--/--/--/--/-/→ Rio de Janeiro

1000 2000 **2690 km**
--/--/--/--/--/--/--/--/--/--/--/→ Southern Thule (78)

1961–3 evacuation due to volcanic eruption

1500 1600 1700 1800 1900 2000
-/-/--/-/

1506 discovered by Tristão da Cunha

7. Nov. 1817 William Glass signs the agreement of communal living

**Edinburgh
of the Seven Seas**

Rookery Point

Hottentot Point

*Boatharbour
Bay*

BIG
GREEN
HILL

West Jew's Point

The Ponds

East Jew's Point

*Runaway
Beach*

*Halfway
Beach*

*THE
HILLPIECE*

The Bluff

Queen Mary's
Peak • 2062
• 1969

Big Gulch

*Sandy
Point*

*Anchorstock
Point*

Mount
Olav

Deep Gulch

RED
HILL

Third Gulch

*GREEN
HILL*

*Noisy
Beach*

Longbluff

*ROUND
HILL*

Tripot Bay

Cave Point

*Seal
Bay Dead-
man's
Bay*

Stonybeach Bay

Stonyhill Point

0   1   2   3   4   5 km
|--|--|--|--|--|--|

Tristan da Cunha

REVOLUTIONS BREAK OUT ON SHIPS, and utopias are lived on islands. It is comforting to think that there must be something more than the here and now. That is the entire thrust of the two books on the shelves of the enlightened German bourgeoisie: the Bible and *Die Insel Felsenburg* (*Felsenburg Island*), a 2,300-page utopian novel about a fictional rocky spot in the wide ocean. But paradise is distant: it may be easier to get to heaven than to an island in the southern Atlantic which is a republic, a country of the just and a model of a better world. The law of this snow dome land is simple and bold. Everyone is equal, everything is shared, and a marvellous patriarch oversees everything. Their happiness is manifested in monogamous marriage. Nine families exchange food; fruit and vines grow in the wild. In the island's interior, there are secret tunnels through caves and a waterfall. And only good men, specially chosen, are allowed in. Anyone bad, or who has evil intentions, will drown in the harsh sea in any case. But whoever who runs aground here and wishes to stay must tell the story of his life as if it were the story of a stranger. Those who have failed in the larger world are always best suited to life in utopia. A new beginning, a fundamentally better life, another 'I' is possible. // *It's off to Felsenburg Island for me*, thinks Arno Schmidt, who believes he has found it in Tristan da Cunha. For it was here, a hundred years after Johann Gottfried Schnabel's novel first appeared, that William Glass lived with his followers in a state of modest Tristanian micro-communism, in just the way that Schnabel had predicted. Arno Schmidt calls for an unabridged version of his favourite book, *Die Insel Felsenburg* to manifest itself and requests a plot of earth on this distant island: *But surely they will have to grant me, who can prove that this strangest of all islands is tremendously interesting, the right to settle there? What about 20 acres then; right next to the small radio station; and a corrugated iron hut, just 80 metres square? – I will pay the money for the voyage there myself.* But Schmidt stays in his Lower Saxony moorland. Grapes never grow on Tristan da Cunha. And Felsenburg Island is still not on the map.

^{59° 27' S}
^{27° 18' W}

Southern Thule

South Sandwich Islands

(United Kingdom)

36 km² | uninhabited

740 km
--/--/--/⟶ South Georgia

1400 km
--/--/--/--/--/-/⟶ Antarctica

960 km
--/--/--/--/⟶ Laurie Island (224)

31. Jan. 1775 discovered by James Cook

1500   1600   1700   1800   1900   2000
-/-/--/--/-/--/-/--/--/--/-/--/--/--/-/--/--/--/--/--/--/--/--/--/--/--/--/-

1976 – 82 occupied by Argentina

Salamander
Point

BELLINGSHAUSEN •255
ISLAND Basilisk
Peak

Beach Point

Tilbrook
Point

THULE
ISLAND
(Morrell
Island)

Mount Harmer
•1115

Resolution
Point

Mount
Larsen
•710

ell
t

Reef
Point

DOUGLAS STRAIT

COOK ISLAND

Twitcher
Rock

Ferguson
Bay

0 1 2 3 4 5 km
|---|---|---|---|---|

Southern Thule

THE ROMANS called the very edge of their flat world *Thule*. So where does it lie? At the outermost of all borders. In the Arctic Circle. Just before the end of the world is nailed down with boards, the last post in the known world is an island in the far north, where the sea is so wild and forbidding that no one wants to travel there, a day's journey away from where the seas flow into each other. // Commander Cook's second voyage takes him south. His mission is to finally find *Terra Australis*, the mighty supposed continent that stretches immeasurably wide across the world map, a huge land mass with a temperate climate, rich in natural resources and with civilized people: a place that is world famous but as yet undiscovered. // In January 1775, his *Resolution* voyages into the Antarctic Ocean for the fourth time. But once again, enormous stretches of pack ice and ice floes force them to turn back; everyone on board is glad when, just a few miles past the 60th parallel, they steer north again. The sailors have had enough of the wet, foggy conditions and the bitter cold, of working the icy masts and rigging, of constant frostbite and rheumatic pains; some are so exhausted they have fallen into day-long faints. // Suddenly they come upon a frozen land with black cliffs, precipitous and full of hollows: cormorants live up high, while unruly waves crash round below. Thick cloud covers its mountains – only one snowy peak rises far above it, looking at least two miles high. Five nautical miles later, they see another mountain, the southern edge of this barren land, perhaps the northernmost tip of the continent they are searching for, which – so much is certain now – will never be of much use: it is a land of firn and ice ruins that never melt, gloomy, cold and full of horrors. Shrouded in thick darkness, they abandon this part of the world to the mercies of Nature. Here is the new Thule, the other end of the known world.

INDIAN
OCEAN

Trome

Possession Island

Diego Garcia

Christmas Island

South Keeling Islands

Amsterdam Island

Saint Paul Island

Saint Paul Island

(France)

FRENCH *Île Saint-Paul*

7 km² | uninhabited

1000   2000   *3010 km*
--/--/--/--/--/--/--/--/--/--/--/--//→ Antarctica

1000   2000   3000   4000 *4290 km*
--/--/--/--/--/--/--/--/--/--/--/--/--/--/--/--//→ South Afri

1000   2000 *2260 km*
--/--/--/--/--/--/--/--/--//→ Possession Island (92)

19.Apr.1618 sighted by Harwick Claesz de Hillegom
1500   160/   1700   1800   1900   2000
-/-
1559 mentioned on a Portuguese map
24.Oct.1892 annexed by France

Pointe Schmith

La Quille

264
Crête
de la
Novara

Pointe Ouest

Pointe Hutchison

Pointe Sud

0 1 2 3 4 5 km
|---|---|---|---|---|

Saint Paul Island

ON 18 JUNE 1871, the English post ship HMS *Megaera* runs aground on one of the natural gravel sea breaks at the entrance to the crater. The shipwrecked crew struggles ashore, and is greeted by two Frenchmen. They come from the island of Bourbon, and do not speak a word of English. // One of the two men calls himself *the governor*. He is thirty years old and lame in one leg. The other man, who introduces himself as *the subject*, is five years younger and in excellent physical condition; he is a splendid climber, for whom no cliff face is too steep. He readily shows the stranded men round the island while the governor crouches in front of a hut at the edge of the crater. The subject refers to him without exception as *a very good man*. The governor unfailingly describes his subject as *a thoroughly bad man*. Never have two people been more suited. They live together in a tiny wooden hut with a small collection of French books. The two of them have been inseparable for an eternity. Their task is to watch over the four small boats lying in the flooded basin of the crater, and to register whalers –

for a monthly wage of forty francs. But almost no one ever lands here, in an area feared for its terrible storms and thick fog. // Ducks, rats and wild cats are the only edible animals that live on the island, and apart from a spinach-like leaf, only moss, ferns and dry grass grow here. Once a year, large colonies of penguins come to lay their eggs in the sparse tufts of grass between the rocks. The enormous birds have white breasts, grey backs, bright pink eyes and golden feathers on their heads. They are not afraid of humans, but their flesh is not good to eat. // It is said that there used to be a mulatto living here with the Frenchmen. The good man and the bad man are said to have murdered him and eaten him up, and to keep all that remains in the very same hut that the governor watches over, day in and day out.

12° 10' S
96° 52' E

South Keeling Islands

(Australia)

Also known as the *South Cocos Islands*

13.1 km² | 596 inhabitants

1110 km
--/--/--/--/-/--→ Java

1000   2100 km
--/--/--/--/--/--/--/--/--/-/--→ Australia

960 km
--/--/--/--/--/--→ Christmas Island (104)

1826–31 battles between the first two settlers,
Alexander Hare and John Clunies-Ross

1500   1600   1700   1800   1900   2000
-/-

1609 thought to be discovered by William Keeling

1978 Australia buys the islands from the Clunies-Ross family

HORSBURGH ISLAND
(Pulo Luar)

Possession
Point

DIRECTION ISLAND
(Pulo Tikus)

PORT REFUGE

Turk Reef

HOME ISLAND

WESTERN ENTRANCE

Pulo Ampang

Ujong Tanjong

Pulo Cheplok

Pulo Pandang

WEST ISLAND
(Pulo Panjang)

LAGOON

Alor Pinyu

Telok
Grongeng

Pulo
Kambing

Ujong Pulo Jau

SOUTH ISLAND
(Pulo Atas)

0 1 2 3 4 5 km
|--|--|--|--|--|--|

South Keeling Islands

HMS *Beagle* drops anchor in the lagoon for twelve days, a gentle body of water lapped by foaming waves and framed by reefs. Charles Darwin explores the islands, collecting specimens. He discovers twenty species, nineteen genera and sixteen families among the plants, all descendants of the stray seeds the sea has carried here. The land consists entirely of rounded coral pieces. It is teeming with hermit crabs; on their backs are mussel shells they have stolen from the neighbouring beach. // On 4 April 1836, the sea is exceptionally calm, so Darwin dares to wade over the outermost barrier of dead rock out to the living walls of coral, against which the waves of the open sea are breaking. Here, between the tide barriers, the expanses of intricate coral flourish; soft, shimmering underwater beings that dry up in the air and in the sunlight. Day and night, the small flower-like organisms face the unstoppable apparent violence of the pounding waves; with their combined strength, they hold out against it. Once they lined the cone of a volcano, dying with it when it sank into the ocean. All that was left of them was a limestone skeleton, on which succeeding generations of coral settled. Remnants of the collapsed mountain came to rest on them, and the sand blown hither by the wind collected here. Slowly, an island grew out of the limestone, the tireless work of the coral – builder and material alike. Every atoll stands as a monument to an island that has gone under, a miracle greater than the pyramids, solely created by these tiny, delicate creatures. // When the *Beagle* leaves the lagoon, Darwin writes, *I am glad we have visited these islands: such formations surely rank high amongst the wonderful objects of this world.* Years later he would come to the conclusion: *The tree of life should perhaps be called the coral of life.*

46° 24' S
51° 45' E

Possession Island

Crozet Islands (France)

FRENCH *Île de la Possession*, originally
Île de la Prise de Possession ['Appropriation Island']

150 km² | 26–45 residents

2150 km
--/--/--/--/--/--/--/--/-/→ Antarctica
1000

2370 km
--/--/--/--/--/--/--/--/--/-/→ Madagascar
1000   2000

3460 km
--/--/--/--/--/--/--/--/--/--/--/--/→ Bouvet Island (70)
1000   2000   3000

24. Jan. 1772 discovered by Marc-Joseph Marion du Fresne
-/-
1500   1600   1700   1800   1900   2000

1964 research station opened

Cap Vertical

Pointe Sombre

Pointe Basse

Cap de la Meurthe

Roche Percée

BAIE DE LA HÉBÉ

•769
MONTS JULES VERNE

Cap de l'Antarès

BAIE
AMÉRICAINE

Les Aiguilles
•671

Moby Dick

Cap Chivaud

de
oine

Mont
des Cratères
848•

Mont
du Mischief
•821

PLATEAU JEANNEL

Pointe Max
Douguet

CIRQUE
AUX MILLE
COULEURS

Baie
du Marin

Port Alfred

Pointe
Lieutard

934•
Pic du
Mascarin

Lac
Perdu

Malpassée

Pointe du
Bougainville

Styx

Baie du
La Pérouse

Cap du Gallieni

Rochers
de la Fortune

Cap du Gauss

0   1   2   3   4   5 km
|--|--|--|--|--|--|

Possession Island

IN 1962, the French name their first mission to the northernmost massif after the greatest engineer of fantasy their country has ever produced. Today, a precipitous mountain range on the island of Possession and a crater on the far side of the moon – both just the kind of places that he might have travelled to on his extraordinary journeys – bear the name of Jules Verne. The man who waxed nostalgic over the future, and prophesied the past, compressed before and after and near and far into spaces that could be travelled through in patented machines as well-upholstered as the stories he told. Verne's novels are the equivalent of a visit to the World Fair, offering a naturally occurring cabinet of potential adventure, polished to a high technological sheen – daydreams for everyday use, atlases for those who stay at home. // His heroes are boys and young men who spend their lives travelling in an attempt to discover the secrets of the world through the acquisition of encyclopaedic knowledge: Dr. Samuel Fergusson, who claims, *I follow no path – the path follows me*, and

Captain Nemo, the lover of the sea. // The journeys to the moon, to the centre of the Earth and to the underworld satisfy both boundless curiosity and the need for security. A few kilometres south of Mont Jules Verne, the river Styx flows from the lost lake into the open sea, which stretches to Antarctica. // This barren archipelago is so difficult to get to, you might think the only way to reach it was to be dragged by the constant drift of the west wind that pushes ships from Africa to Australia, dashing so many to pieces on this island's jagged cliffs, and then wrecked against its scattered basalt rocks. // But Jules Verne's mysterious island is far away, somewhere in the Pacific Ocean, and this is a most inhospitable place for any aspiring Robinson Crusoes.

7° 18′ S
72° 24′ E
Diego Garcia *Chagos Archipelago* (United Kingdom)

27 km² | 3,700 residents

780 km
--/--/--/→ Maldives

1000   *1780 km*
--/--/--/--/--/--/--/--/→ India

1000   *2000*   *2710 km*
--/--/--/--/--/--/--/--/--/--/--/--/→ South Keeling Islands (88)

1967–73 resettlement of the Chagossians

1500   *1600*   *1700*   *1800*   *1900*   *200(*
-/-
/
After 1500 discovered by Portuguese sea farers

Since 2000 legal battle over the right of return

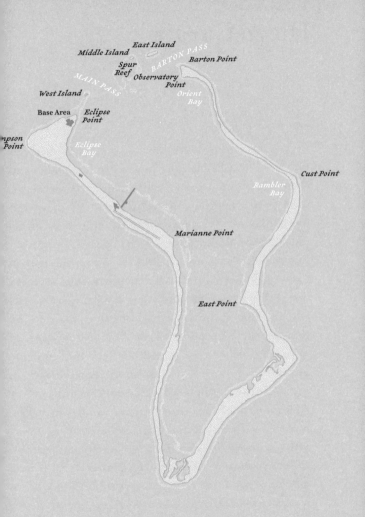

East Island

Middle Island

Spur
Reef

West Island

BARTON PASS

Barton Point

Observatory
Point

Orient
Bay

MAIN PASS

Base Area

Eclipse
Point

mpson
Point

Eclipse
Bay

Cust Point

Rambler
Bay

Marianne Point

East Point

0   1   2   3   4   5 km
|---|---|---|---|---|

Diego Garcia

IN THE SLUMS OF PORT LOUIS, they are waiting to return. The Chagossians lost their homeland and their life in a modest paradise over forty years ago. They are a people not allowed to be one, for then what has happened to them would be an injustice, a criminal act by a colonial power, a dirty deal in the glittering ocean. For three million pounds, the British Crown granted Mauritius independence and retained the archipelago. For one dollar a year, it leases the islands – for an initial period of fifty years – to the greatest of all countries in the brotherhood of nations. // Now there is a military base in the middle of the Indian Ocean. It is the most secretive base in the world, but it pitches itself as a dream destination: *This spectacular area east of equatorial Africa – where a 30-minute bus tour can show you the entire location – holds elements of an adventure vacationer's dream. There's tropical windsurfing and fishing for 200-pound marlin. And the sea is so warm, snorklers can wade in and play tourist with thousands of brilliantly coloured tropical fish. Apart from clubs and a golf course, the base has a sports hall, a gallery, a shop, a library, a post office, two banks and a church. Our motto is: One Island, One Team, One Mission.* No one mentions the five hundred families that were forcibly deported and declared to be immigrant workers. Instead, British diplomats assure the world that the islands were previously uninhabited. // The atoll takes the form of two fingers spread in a crooked V, a victory sign in the Indian Ocean. But victory for whom? The Chagossians fight for British passports, access to the courts and ultimately for the right of return. But it is taken away from them again. The Queen signs an agreement, another relic from colonial times; and the Chagossians' homeland remains a restricted zone, a navy and air force base. Its name: *Camp Justice.*

³⁷° ⁵⁰' ˢ
⁷⁷° ³³' ᴱ

Amsterdam Island

(France)

FRENCH *Nouvelle Amsterdam* ['New Amsterdam']

58 km² | 25 residents

--|--|--|--|--|--|--|--|--|--|--|--|--|--|--|--|--|--|--/--//→ South Afri⟩
  1000   2000   3000   4000 *4290 km*

--|--|--|--|--|--|--|--|--|--|--|--|--|--|--|--/--/→ Australia
  1000   2000   3000   72° 24' E

90 km
-/→ Saint Paul Island (84)

18. Mar. 1522 sighted by Juan Sebastián de Elcano
        /
                        1949 construction of meteorological station⟩
                        /
1500/   *1600*   *1700*   *1800*   *1900* /   *2000*
-/--|--|--|--|--|--|--|--|--|--|--|--|--|--|--|--|--|--|--/-

June 1633: named by Anthonie van Diemen
   after his ship *Nieuw Amsterdam*

                Dec. 1997 to Feb. 1998 Alfred van Cleef
                        stay on the island

Pointe
Goodenough

Base Martin de Viviès

Pointe de
la Recherche Cratère Antonnelli
●202
Cratère Vénus
●303

Pointe de
l'Eboulement

Cratère de l'Olympe
●691

Mont
Fernand
●731
●742 La
Grande Marmite
●881
Mont de la Dives

Pointe de
la Novara

Fausse Pointe

Falaise
d'Entrecasteaux

LES
GRANDES
RAVINES

Pointe del Cano

Pointe
Vlaming

0 1 2 3 4 5 km
|---|---|---|---|---|

Amsterdam Island

NO ONE IS ALLOWED to settle here, so the personnel at the research station changes constantly. Some of the men stay for only a few months, but most for a year and a half, on the island that they call simply *Ams* or *La Base*. None speak English and all greet each other every day with a handshake. There is no boat. Where would they take it? This place is a stray piece of France, a cross on a blue nowhere on the various maps of the world that are pinned to the walls, alongside a couple of pictures of albatrosses and countless pornographic posters. // Domestic cattle gone wild graze in a large enclosure fenced off with barbed wire. Down on the beach, male fur seals howl. The sea mammals heave themselves ponderously over the rocks. The bulls are fighting for the females, due to arrive in a few days. The victors take up the best spots by the sea. // In the dining hall of *Great Skua*, the district chief of the 48th mission gives a speech after dinner: *There is no such thing as isolation. Even on Amsterdam Island, we are cogs in a huge wheel; here too, we receive signals that tell us who we are.* He calls himself a fantasist, a doctor and a professional soldier, in that order. His office is the only room without pin-ups on the wall. There is a register of births, marriages and deaths on his desk. Empty columns show that no one has married or had a child here yet. Everyone who stays on Amsterdam for longer than a year is examined by a medical officer from the south of France to check that he is coping with the long period of restriction of movement and the confined, purely masculine environment. No woman has visited for longer than two days. // At night, the men gather in the small video room in *Great Skua* to watch one of the porn films from their personal collection. Each man sits in a row on his own. The loudspeakers emit grunts and groans, and the air is heavy with the musky scent of the bull seals.

10° 30' S
105° 38' E

Christmas Island

(Australia)

135 km² | 1,402 inhabitants

350 km
--/-/→ Java

1000　　　　*2000*　*2590 km*
--/--/--/--/--/--/--/--/--/--/--/--/→ Perth

1000　　　　*2000*　　　*3000*　　*4120 km*
--/--/--/--/--/--/--/--/--/--/--/--/--/--/--/--/···/→ Amsterdam
Island
(100)

25.Dec.1643 discovered by William Mynors

1500　　*1600*　　*1700*　　*1800*　　*1900*　　*2000*
-/-

1989 discovery of the first supercolony of crazy ants

Northeast Point

Rocky Point

Silver City

The Settlement

Poor Saan

Smith Point

Drumsite

Northwest Point

Hanitch Hill
309

Waterfall

TOM'S RIDGE

Vincent Point

NORTHERN PLATEAU

Martin Point

Murray Hill   349
361   Jacks Hill
318
Ferguson Hill

Wright Point

Deans Point

Toms Point

Ross Hill
319

John D. Point

Middle Point

Jones Point

SMITHSON BIGHT

SOUTHERN PLATEAU

Tait Point

Stubbings Point

Medwin Point

South Point

0 1 2 3 4   5 km
|--|--|--|--|--|--|--|--|--|--|

Christmas Island

THE RAINY SEASON lures them out of their holes. Every November, 120 million crabs reach sexual maturity and make their way to the sea. A red carpet spreads over the island as they crawl over tarmac and thresholds and climb over walls and rock faces, thrusting their shells sideways on their two strong pincers and eight thin legs, heading for the ocean to cast their black spawn into the waves just before the new moon. // Not all reach their destination. Their enemies are everywhere. No one knows exactly where they came from. The yellow crazy ants simply appeared one day, brought in by visitors. The invaders are only four millimetres long, but an army of them is a destructive force. The ant populations coexist peacefully and their queens have made a fateful pact: together they form united colonies; a superpower, an empire. Behind each of the three hundred queens follows a gigantic army of workers, with long, bendy legs, slender yellow trunks and brownish heads. // They build nests in hollow trees and deep cracks in the earth and keep scale insects to produce the sweet honeydew they feed on. They move at lightning speed, changing direction every few seconds, and are always ready to attack. Their victims are the booby and frigate bird hatchlings in their nests and the red crabs migrating to sea. The crazy ants spray formic acid on to their fiery shells. First the crabs are blinded, then they lose their bright colour, and three days later they are dead. It is war on Christmas Island.

15° 53' S
54° 31' E

Tromelin

Scattered Islands (France)

FRENCH formally *Île des Sables* ['Sand Island']

0.8 km² | 4 residents

430 km
--/-/→ Madagascar

550 km
--/--//→ Mauritius

1000   2160 km
--/--/--/--/--/--/--/--/-/→ Diego Garcia (96)

                              31.Jul.1761 shipwreck of the *Utile*
  1500   1600   1700 /   1800   1900   2000
-/--/--/--/--/--/--/--/--/--/--/--/--/-/--/--/--/--/-/--/--/--/-
                              /
1722 discovered by Briand de la Feuillée

Station météo

*Barrière
des récifs*

0   1   2   3   4   5 km
|---/--/---/---/---/

Tromelin

ON 17 NOVEMBER 1760, the *Utile*, an East India Company ship, leaves Bayonne in south-west France for the Mascarene Islands. The vessel stops in Madagascar to replenish food supplies and, against the governor's orders, the captain, Jean de la Fargue, takes sixty slaves on board to sell along with the other goods when they reach Île de France (modern-day Mauritius). But when a storm throws the *Utile* off course, she runs aground and is smashed to pieces on the reef of this small island. It is a strip of sand with a couple of palm trees, barely two kilometres long and eight hundred metres wide, simply called Sand Island. Almost everyone who makes it ashore is badly injured, more dead than alive. // The survivors begin to build a boat out of the wreckage of the ship. Two months later, it is ready. The French sailors board, all 122 of them, packed in tightly, promising to fetch help, and are never seen again. The slaves stay behind. They are free, but trapped as never before, slaves now to their desire to survive. They make a fire, dig a well, sew themselves clothes out of feathers and catch seabirds, turtles and shellfish. Some are so desperate that they float off into the unknown on a raft – anything is better than to stay trapped on a patch of sand, at the mercy of hope and their own lives. The others stay watching the fire. It is still burning after fifteen years. Of the sixty former slaves, just seven women and a baby, not yet weaned, remain. On 29 November 1776, the crew of the corvette *Le Dauphin* takes them on board and brings them to Île de France. They leave nothing behind on Sand Island apart from the charred wood of their dead fire and the name of their saviour, an officer of the French royal navy, the captain of the corvette: the Chevalier de Tromelin.

PACIFIC
OCEAN

St George Island

Semisopochnoi

Atlasov Island

Iwo Jima

Pagan

Taongi

Pingelap

Howland Island

Banaba

Takuu

Pukapuka

Tikopia

Napuka

Fangataufa

Rapa

Norfolk Island

Raoul Island

Antipodes Island

Macquarie Island

Campbell Island

Socorro Island

Clipperton Atoll

Cocos Island

Floreana

Pitcairn Island

Easter Island

Robinson Crusoe

14° 10' S
141° 14' W

Napuka *Disappointment Islands* (French Polynesia)

Also known as *Pukaroa*, formerly *Wytoohee*

8 km² | 277 inhabitants

20 km
/→ Tepoto Nord

1000   2000   3000   3990 km
--/--/--/--/--/--/--/--/--/--/--/--/--/--/--/--/--/→ Hawaii

920 km
--/--/--/-/→ Fangataufa (134)

1977 opening of the airport

1500   1600   1700   1800   1900   2000
-/--/--/--/--/--/--/--/--/--/--/--/--/--/--/--/--/--/--/-

End Jan. 1521 thought to be discovered by Ferdinand Magellan

Rangihoa
Titika
Onamu
Kavake

Oire
KOKO
Manga
Manga

Tupiti
Karena
Maihiva

O Homo
Ngake

Onimo
Araveke
Kurima

Mirinuku
Ongare

0 1 2 3 4 5 km

Napuka

WHEN THEY REACH the great ocean on 28 November 1520 and set course for the north-west, the captain-general Ferdinand Magellan announces that they will take a month at most to reach the Spice Islands. But no one believes that any more. They have seen no land for weeks. The ocean is perfectly calm, and they name it for its stillness: *Mare Pacifico*. It is as if the gates to eternity have opened, and they are steering directly through them. Soon the compass needle will not have enough strength to point north, and the crew will not have enough to eat. The ship's biscuit is little more than dust, infested with maggots and mouse droppings, and the drinking water is a stinking yellow liquid. Starving, they eat sawdust and the strips wound round the yardarm to protect the ropes. They soak the rock-hard leather in the sea for four or five days to soften it, fry it on the coals and force it down their throats. // When they discover rats, the hunt begins. Up to half a ducat is paid per half-starved specimen. One man is so impatient that he wolfs down his purchase raw, and two sailors get into such an enormous fight over a rat they have caught that one of them strikes the other dead with an axe. The murderer is to be quartered, but no one has the strength to carry out the sentence, so they strangle him and throw him overboard. // Every time someone dies, Magellan makes haste to get the corpse sewn into a sailcloth and tossed into the ocean before his men can turn into cannibals. The survivors stare at the newly dead with increasingly greedy eyes. // When they finally sight land after fifty days, they find nowhere to drop anchor, and the boats that they land on the island discover nothing to satisfy either hunger or thirst. They name the islands *the Disappointment Islands*, and continue their journey. The keeper of the ship's journal, Antonio Pigafetta, writes, *And I believe nevermore will any man undertake to make such a voyage.*

27° 36' S
144° 20' W

Rapa Iti *Austral Islands*
(French Polynesia)

Also known as *Rapa,* formerly *Oparo Island*

40 km² | 482 inhabitants

1180 km
--/--/--/--/-/⟶ Tahiti

```
              1000              2000              3000        3620 km
--/--/--/--/--/--/--/--/--/--/-/--/--/--/--/--/--/-/⟶ New Zealand
```

1000　　1440 km
--/--/--/--/--/--/⟶ Pitcairn Island (178)

1791 sighted by George Vancouver

```
  1500      1600      1700      |1800      1900      2000
-/--/--/--/--/--/--/--/--/--/--/--/--/--/--/--/--/--/--/--/--/-
```

26. *May 1998* Marc Liblin dies on Rapa Iti, aged fifty

Auroa
Point

*Angairao
Bay*

Matapu Point

Autea Point

Mount Vairu
● 218 Mount
Pukunia
● 246

Mount
Perahu
● 385

Atanui Bay

Area

Nukutere Point

Anarua Bay

● 284
Mount Motu

AHUREI BAY

Maomao Point

HIRI BAY

Ahurei

Mount Pukumaru
● 355

Anatauri Bay

Tauturau Island

0   1   2   3   4   5 km
|---|---|---|---|---|---|

Rapa Iti

IN A SMALL TOWN in the foothills of the Vosges, a six-year-old boy is visited by dreams in which he is taught a completely unknown language. Little Marc Liblin soon speaks this language fluently without knowing where it comes from or whether it even really exists. // He is a gifted but lonely child, with a thirst for knowledge. In his youth, he lives on books rather than on bread. At the age of thirty-three, he is an outsider living on the fringes of society when he comes to the attention of researchers from the University of Rennes. They want to decipher and translate his language. For two years, they feed the strange sounds he makes into giant computers. In vain. // Eventually, they decide to trawl through the bars by the harbour to see if any of the sailors on shore leave have ever heard the language before. In a bar in Rennes, Marc Liblin gives a solo performance, holding a monologue in front of a group of Tunisians. The barkeeper, a former navy man, interrupts and says he has heard this tongue before, on one of the most remote Polynesian islands. And he knows an elderly lady who speaks it; she is divorced from an army officer and now lives in a council estate in the suburbs. // The meeting with the Polynesian woman changes Liblin's life. When Meretuini Make opens the door, Marc greets her in his language, and she answers straight away in the old Rapa of her homeland. // Marc Liblin, who has never been outside Europe, marries the only woman who understands him, and in 1983 he leaves with her for the island where his language is spoken.

Robinson Crusoe

Juan Fernándes Islands (Chile)

SPANISH *Isla Robinsón Crusoe*, formerly *Isla Más a Tierra*

['The Island Closer to Land']

96.4 km² | 633 inhabitants

150 km
-/→ Alexander Selkirk Island

630 km
--/--/-/→ Chile

1000   2000   3000   3770 km
--/--/--/--/--/--/--/--/--/--/--/--/--/--/--/--//→ Floreana (158)

1500   1600   1700   1800   1900   2000
-/-

1704–8 Alexander Selkirk's 'Robinson Crusoe' adventure

Punta Norte

Punta
Suroeste

Cerro Alto
600 •

Punta Salinas

*Puerto
Inglés*

Cerro
Agudo
• 685

• 720
Cerro
Portezuelo

San Juan Bautista

Bahía Cumberland

*B A H Í A
T R E S
P U N T A S*

Punta Lemos

CORDÓN ESCARPADO

Cerro
El Yunque
• 915

*Bahía
Villagra*

Cerro
Damajuana
• 635

Cerro
La Piña
• 604

Punta
Tunquillax

hía Padre

Punta
Isla

Islote
Viñilla

Punta
O'Higgins

Punta
Hueca

Playa Larga

*Puerto
Francés*

Punta
Hueso Ballena

ISLA SANTA CLARA

Punta
Freddy

0   1   2   3   4   5 km
|---|---|---|---|---|---|

Robinson Crusoe

ROBINSON CRUSOE's diary is in Berlin, lying on a forgotten shelf in the State Library of Prussian Cultural Heritage. So claims David Caldwell of the National Museum of Scotland. The library is busy: the same faces have been coming here for a decade. There they are behind the encyclopaedias on the top floor, beneath the terrace with its globes as tall as men. Each desk is an island kingdom. They all come here to write: a page a day when things are going well; half a sentence when things are not. // Caldwell spent a month on this archipelago. All he found was an angular, pointed piece of bronze, 1.6 centimetres long. He is certain it must have been part of Alexander Selkirk's dividers, from his navigation equipment. // The diary that the stranded pirate wrote in his solitude ended up in the collection of the Duke of Hamilton but was later auctioned to the nascent German empire. The first novel in the English language was based on it. The published confection has as much imagination as truth in it: Alexander becomes Robinson; the Scottish son of a cobbler becomes a merchant's son from York who ignores the advice of his father; four years and four months becomes twenty-eight years, half a lifetime. The pirate Selkirk becomes the plantation owner Crusoe who constantly struggles with a restless desire to travel to distant lands, but as soon as he achieves this, yearns for his homeland. // There is an occasional rustling in the magazine section and in the evening, when the rows are lit up, the blinds at the glass front of the library are whipped round as if in a little dance, fragmenting the panoramic view on to an empty square. In the manuscripts section, they are sifting through the inventory. On 4 February 2009, a spokeswoman announces, *We have searched all the relevant catalogues and have not found what we were looking for. It is almost certain that Selkirk's diary is not in our collection.* Writers have it much easier than archaeologists.

0° 48' N
176° 37' W
Howland Island
Phoenix Islands (United States)

1.84 km² | uninhabited

1000   *1640 km*
--/--/--/--/--/--/-/⟶ Samoa

1000   2000   *3030 km*
--/--/--/--/--/--/--/--/--/--/--/--//⟶ Hawaii

1000   *1750 km*
--/--/--/--/--/--/--//⟶ Pukapuka (150)

1.Dec.1828 discovered by Daniel McKenzie

1500   1600   1700   1800   1900   2000
-/-/--/-/--/-/--/-/--/-/--/-/--/-/--/-/--/-/--/-/--/-/--/-/--/-/--/-/-

Earhart
Light

0   1   2   3   4   5 km
|---|---|---|---|---|

Howland Island

SHE IS THE FIRST WOMAN to fly solo across the Atlantic – from Newfoundland to Northern Ireland in fourteen hours and fifty-six minutes – and only the second person to do so, after Lindbergh. She flies from Los Angeles to New Jersey, from Mexico City to Newark and from Honolulu to Oakland: Amelia Earhart, a pioneer who writes her records in the sky with vapour trails. Her greatest successes are at high altitude. Over and over again, she is the first woman. But now she wants to attempt something that no one has ever done: she wants to be the first person to circumnavigate the Earth at the equator. *Please know I am quite aware of the hazards*. The last picture taken before her 29,000-mile journey shows an incongruous pair in front of her Lockheed L-10E Electra, a streamlined silver twin-engine propeller plane. Amelia Earhart's hands are resting on her hips. Her flight suit is zipped wide open, her curly-haired head tipped to one side, and she is grinning in a daredevil fashion. She is tall and thin. Next to her, standing like a shy, diligent girl, is her navigator, Fred Noonan. // On the morning of 2

July 1937, the aircraft shoots over the bumpy strip at Lae on the edge of the Solomon Sea, laden with a full tank which has enough fuel for a little more than twenty hours. The whole world – 22,000 miles – is behind them; they just have to complete the last section, over the silent ocean that spreads over half the Earth. // Off Howland Island – 2,556 miles away – the *Itasca*, an American Coast Guard cutter, is waiting for them with fuel and freshly made beds. The atoll is so small that a single cloud is enough to obscure it from view. At 7.42 a.m., Earhart's voice is heard on the radio: *We must be on you, but cannot see you – fuel is running low*. One hour later comes a new call: *We are on the line 157–337, we are running on line north and south*. The crew of the *Itasca* search the horizon with binoculars and send signals, but there is no reply. Amelia Earhart disappears just beyond the date line on a flight into yesterday. The ocean is silent.

54° 38′ S
158° 52′ E

Macquarie Island
(Australia)

128 km² | 20–40 residents

1070 km
--/--/--/--/-/⟶ New Zealand

1000   1510 km
--/--/--/--/--/--// ⟶ Antarctica

700 km
--/--/--/ ⟶ Campbell Island (166)

25. May 1948 research station is opened

1500   1600   1700   1800   1900   2000
-/-

11. Jul. 1810 discovered by Frederick Hasselborough

North Head

Hasselborough
Bay

Handspike Point

Anare
Station

Halfmoon Bay

Buckles
Bay

Mount
Elder
•371

Nuggets Point

Langdon Point

Douglas Bay

347 •
Mount
Power

Tussock
Point

Bauer
Bay

Mawson Point

Cormorant Point

Sandy
Bay

Brothers
Point

Mount Eitel
•361

Aurora Point

Mount
Ifould
374 •

Mount Law
•347

Green Gorge

Mount
Waite
•422

Double
Point

Davis
Bay

Mount Blake
•372

Sandell
Bay

Saddle Point

Waterfall
Bay

Cape
Toucher

•433 Mount
Hamilton

Mount
Fletcher
•428

Lusitania
Bay

Precarious
Point

Cape Star

Carrick
Bay

Caroline Point

Mount
Ainsworth
•363

Windsor
Bay

Hurd
Point

0 1 2 3 4 5 km
|---|---|---|---|---|

Macquarie Island

THIS CRAGGY PIECE of land, where it rains all year round, has never been part of a land mass, but comes directly from the sea. It is a piece of the Earth's crust from the bottom of the ocean that just happened to shoot up above sea level, a vertebra of an undersea spine that rises above the water. Here, halfway to Antarctica, where the warm water of the north meets the cold water of the south, the sea is always stormy, and every landing is dangerous. // It is only with great difficulty that, in January 1840, the crew of the *Peacock* manage to land without losing their ship. The men explore the steep rocky ground, gathering specimens of the sparse vegetation. Lieutenant Charles Wilkes comes to the conclusion that *Macquarie Island offers no inducement for a visit.* // Only the midshipman Henry Eld is overwhelmed when he walks down to Hurd Point on his own. Grass-covered shipwrecks moulder in every bay and on every beach; skeletons of ships in a sea of penguins, of which the island has millions. He has often heard of the great quantity of birds on uninhabited islands, but he is not prepared for this huge number. The whole sides of the rugged hills are covered with them. He has never heard such terrible squeaking, squalling and gabbling before, and he has never dreamed that any of the feathered tribe would be able to make such a din. They snap at him from all directions, catching hold of his trousers, shaking and pinching his flesh violently so that he flinches and stands upon the defensive. With pale bellies and dark faces, beaks stretched out, they surround the intruder. More and more birds come closer, upright and imperturbable, with the dignified tread of strict headmasters, until Henry Eld completely disappears into the field of black and white.

22° 15' S
138° 45' W

Fangataufa *Tuamoto Archipelago* (French Polynesia)

Formerly *Cockburn Island*

5 km² | uninhabited

40 km
/→ Moruroa

--/ /→ New Zea▶
　　　　　　1000　　　　*2000*　　　*3000*　　　*4000* *4410 km*

810 km
--/--/--// → Rapa Iti (118)

Feb. 1826 discovered by Frederick William Beechey
-/--/
　1500　　　*1600*　　　*1700*　　　*1800*　　　*1900*　　　*2000*

1966–96 test site for nuclear weapons

Empereur

Kilo

Pingouin

PASSE BALISÉE

Frégate

Pavillon

Fox

LAGON

Hélène

Echo

Therme
Nord

Therme Sud

0 1 2 3 4 5 km
|---|---|---|---|---|

Fangataufa

THE COLONIES have all been parcelled out and two world wars have been won. Now, to be a world power, you need the Bomb. And the fourth victorious power wants to join in. It wants validity on the world stage; to inspire fear and to prove its might with nuclear power. France's first atom bombs are detonated in the Sahara. When Algeria and its deserts become independent, a new wasteland must be found for the *force de frappe*. At first they consider the isolated Clipperton Atoll, then the wind-buffeted Kerguelen Archipelago. For the terrible deed they finally choose a picturesque place – two lagoon islands in Tuamoto, far away from the eyes of the world: Moruroa and Fangataufa, uninhabited atolls where nature is luxuriant and untouched. // When the French land on Fangataufa, they blast a hole in the northern section of the ring of land in order to make it passable for ships. They distribute protective goggles and sunglasses to the inhabitants of the neighbouring atolls. // On 24 August 1968, everything is ready for the big test: the detonation of the first French hydrogen bomb. It is the largest that has ever been exploded, with a force of 2.6 megatonnes – between a hundred and a thousand times more powerful than an atom bomb. A helium balloon carrying the three-tonne bomb rises to 520 metres. Someone says the code-name for this operation: *Canopus*, the name of the second brightest star in the night sky. The star is so far south that it cannot be seen from France, just like the explosion of this bomb, which occurs at 19.30 Paris time. A gigantic cloud with a coiled tail of exploded steam spreads into the sky. The shock waves extend outwards, casting ringed shadows on the lagoon, the atoll and the sea, pushing the ocean to flood towards the horizon. // Afterwards, nothing remains. No houses, no installation, no trees, nothing. The entire island is evacuated because of radio-active contamination. No one is allowed to set foot on Fangataufa for six years.

50° 51' N
155° 33' E

Atlasov Island

North Kuril Islands (Russia)

RUSSIAN *Ostrov Atlasova*
JAPANESE *Araido-tō*

119 km² | uninhabited

70 km
-/→ Kamchatka

1000   1370 km
--/--/--/--/--/-/→ Sapporo

1000   1650 km
--/--/--/--/--/--/-/→ Semisopochnoi (182)

Early 1950s women's penal colony set up

1500   1600   1700   1800   1900   2000
-/-

Mys
Borodavka

Mys
Rouniy

Buchta
Severnaya

Mys
letscho
Saliv
Otvagi

Alaid

Glawnij

Mys
Chitriy

956

Owraschnij

Mys
Serditiy

Wulkan Alaid
2339
2291
Pik Bonovoy

Pik Glawniy
Gora Parasit
1023

Sapertyj

BUCHTA
BAKLAN

Poluostrov
Vladimira
BUCHTA
ALAIDSKAYA

Gora Osobaya
208

Mys
Podgorniy

Ochotskij

Mys
Siandriom

Mys
Lava

Mys
Devjatka

Mys
Pologiy

0 1 2 3 4 5 km
|--|--|--|--|--|--|

Atlasov Island

THIS ISLAND does not take its name from the Titan who carried the heavens but from a Cossack. It is nothing but a single lonely mountain that is higher than all the other pearls in this chain of islands whose black sands rise above the waters. // The mountain, which the people of the Kuril call *Alaid*, is more beautiful than Mount Fuji. In winter, its peak of grey basalt is covered with sugar-white snow. The volcano is the northernmost of the scattered islands that appeared in a ring of fire 40,000 to 50,000 years ago. Its beauty lies in its symmetrical form. // It is said that it once stood in the middle of Lake Kurile in Kamchatka, towering so high in the sky that it blocked the light from the neighbouring mountains, who were furious and spoiling for a fight. In reality, they were simply jealous of its perfect beauty. This disturbed the mountain greatly, and it felt forced to give up its place in their midst, so it started out on a long journey, finally setting itself down in a peaceful spot far away in the sea. // But in memory of its time in Lake Kurile, and as a sign of mourning, it left behind its heart, which is called *Outchitchi* in the Kuril language, or 'heart of the mountain' in Russian, a cone-shaped rock in the middle of the lake. // The river Ozernaya flows in the tracks of the mountain's reluctant journey. When the mountain lifted itself from its place, the water of the lake rushed after it. It is a thin blue umbilical cord that will always bind the exiled mountain to its homeland.

14° 38' N
169° 0' E

Taongi Atoll

Ratak Chain (Marshall Islands)

Also known as *Bokak*, formerly *Gaspar Rico* and *Smyth Island*

3.2 km² | uninhabited

280 km
--//→ Bikar

1000   *2000*   *3000*   *3750 km*
--/--/--/--/--/--/--/--/--/--/--/--/--/--/--//→ Hawaii

1000   *2000*   *2500 km*
--/--/--/--/--/--/--/--/--/--/--//→ Pagan (210)

21.Aug.1526 discovered by Alonso de Salazar

1500   *1600*   *1700*   *1800*   *1900*   *200*
-/-

10.Sept.1988 the *Sarah Joe* is foun

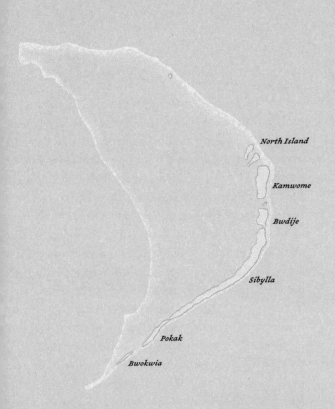

North Island

Kamwome

Bwdije

Sibylla

Pokak

Bwokwia

0 1 2 3 4 5 km
|---|---|---|---|---|

Taongi Atoll

SCOTT MOORMAN grows up in the San Fernando Valley. He watches the TV series *Adventures in Paradise* as a child and dreams of living in Hawaii. In 1975, he leaves the mainland and finds a new home in Nahiku, on the east coast of Maui, which lives by Hawaiian time: when the weather is fine, no work is done. So it is on Sunday morning, 11 February 1979. The ocean is as smooth as glass and there is barely a cloud in the sky. Scott and four friends decide to go fishing. They buy new spark plugs for the motorboat, beer and soft drinks for the fridge and ice for the fish that they hope to catch. At about ten o'clock they pass the rocky island at the mouth of the bay, and steer the five-metre-long *Sarah Joe* southwards. They have long hair and bushy moustaches, and are wearing sunglasses. One of them is rolling the first joint of the day. // Just before noon, the wind rises, turning into a storm by the afternoon and a hurricane over the island by the evening, whipping the sea up and laying waste to the coastline. The waves are several metres high and the rain is relentless. // At five o'clock in the afternoon, the *Sarah Joe* is reported missing. The coastguard sends a helicopter and a plane out in the storm, but visibility is too poor. They extend the search area every day. The coastguard searches for five days, and the men's family and friends continue searching for another week. They find nothing. Nothing at all. Not a trace of the men, not one piece of the boat. // Nine and a half years later, one of the searchers, the marine biologist John Naughton, finds a wrecked boat on the beach of Taongi, the northernmost and driest atoll of the Marshall Islands, 3,750 kilometres west of Hawaii. A Hawaiian registration number is prominently displayed on the fibreglass hull. It is the *Sarah Joe*. // There is a simple grave nearby: a cross of driftwood on a pile of stones. A few bones protrude from the sand. These are discovered to be the remains of Scott Moorman. Who buried him here and where the other men are remains a mystery.

Norfolk Island

(Australia)

NORFUK *Norfuk Ailen*

34.6 km² | 2,128 inhabitants

740 km
--/--/--/→ New Zealand

1000   1390 km
--/--/--/--/--/-/→ Australia

1000   1850 km
--/--/--/--/--/--/-/→ Tikopia (206)

10. Oct. 1774 discovered by James Cook

1500   1600   1700   1800   1900   200
-/-

1788–1813 first penal colony

1825–55 second penal colony

Point
Vincent
Point Howe
Duncombe Bay
Bird Rock
Anson Point
Mount Bates
318 •
316 • Mount Pitt
CASCADE
BAY
Anson
Bay
Puppy's
Point
Cascade
Steels
Point
Burnt Pine
Middlegate
Point Blackbourne
Rocky
Point
Ball Bay
Kingston
Collins Head
Point
Ross
Sydney
Bay
Cemetery
Bay
Nepean
Island

PHILIP
ISLAND

0   1   2   3   4   5 km
|---|---|---|---|---|

Norfolk Island

EXILE TO THIS PARADISE is the harshest punishment for a criminal. No one returns from this hell. The prisoners keep their gaze lowered and speak without moving their lips. They work in opencast mines or on the coastal reefs, where they hack limestone away from the walls of coral. Even the hardest work is better than solitary confinement. Lunch consists of potato and cornmeal mush with salt meat, tough as leather, and water from a bucket. In the evening those who still show even a flicker of resistance are flogged with the cat-o'-nine-tails until they lose consciousness. // Monday, 25 May 1840: it is the Queen's twenty-first birthday. The ships in harbour salute her with their cannons: one blast for each year. The new superintendent, Alexander Maconochie, who has only been in office for a few weeks, announces a celebration: freedom of movement for all! Both prisoners and guards can barely believe it. There are no chains and no security measures. All the gates are opened. Together they drink to the health of the distant queen with punch laced with a few drops of real rum. // The superintendent walks through the open dungeons and the convicts stroll over the hills and wander through the pine woods. In the evening, they dine together al fresco. There is fresh pork, fireworks, and entertainments, which the prisoners have rehearsed. The tent scene from *Richard III* is performed. A prisoner dances a hornpipe with childlike high spirits and another sings the best-loved aria from *The Castle of Andalusia*, the song of the wolf: *While the wolf, in nightly prowl, / Bays the moon, with hideous howl, / Gates are barr'd, a vain resistance! / Females shriek; but no assistance. / Silence, or you meet your fate; / Your keys, your jewels, cash and plate; / Locks, bolts, bars, soon fly asunder, / Then to rifle, rob, and plunder.* // After the national anthem, the horn sounds the signal. Everyone returns to their dungeons and barracks. There are no incidents of any sort to report on this day.

10° 53' S
165° 51' W

Pukapuka
(Cook Islands)

Also known as *Danger Island*

3 km² | 600 inhabitants

700 km
--/--/--/→ Samoa

1000 1300 km
--/--/--/--/--/→ Rarotonga

1000   2000   2680 km
--/--/--/--/--/--/--/--/--/--/--/-/→ Napuka (114)

21.Jun.1765 sighted by John Byron

1500   1600   1700   1800   1900   2000
-/-

20.Aug.1595 discovered by Alvaro de Mendaña

1924 Robert Dean Frisbie moves to Pukapuka

Roto   *Pukapuka*

Yato   Ngake

TE AVA O TE MARIKA   Te Motu o te Mako
(PASSAGE)   *Te Aua Loa*

Te Alai   Nuku Wetau
Motumotu

Toka

Motu Kotawa   *Te Alo*
i Ko   Matau Tu

Matauea

Motu Ko

0   1   2   3   4   5 km

Pukapuka

ROBERT DEAN FRISBIE sits on the veranda of the Pukapuka trading station. Behind him is half the village, in front of him a small collection of scattered huts on the beach. Children are playing in the shallow water and old women are weaving hats out of pandan leaves in the gentle evening breeze. // Suddenly a neighbour runs up to him, completely naked, wet from bathing, her hair plastered to her golden skin. She is out of breath, and her breasts rise and fall as she hurriedly asks for a bottle of something. Frisbie quickly passes her what she wants, and as she disappears into the dusk he stares after her for a long time, strangely moved. Although he has lived here for years now, he has still not got used to the nudity. In this respect, he is still the boy from Cleveland who could never have imagined the freedom here. // On Pukapuka, no one cares if a bride is a virgin. A word for 'virgin' does not even exist in their language. A woman who has a child out of wedlock earns respect and increases her chances of marriage because she has proved her fertility. The young people of the three villages meet at the far edge of the beach once darkness falls. There they fight, dance, sing and sleep with each other. It is common for more than two people to get together. Sex is a game, and jealousy has no place. Singing has its role both in foreplay and post-coitus, but the generations see it differently. The older women think it belongs both before and after sex, while the younger ones think that it should come only after. They agree that no singing should take place during coitus. After sex, the men and women bathe in the sea together. // In these matters, Pukapuka certainly has the edge over Cleveland, Robert Dean Frisbie thinks, as he puts out his veranda light.

49° 41' S
178° 46' E

Antipodes Island
(New Zealand)

Formerly *Penantipode Isle*

21 km² | uninhabited

740 km
--/--/--/→ New Zealand

1000   2000 2370 km
--/--/--/--/--/--/--/--/--/-/→ Antarctica

1000   2000 2270 km
--/--/--/--/--/--/--/--/--//→ Raoul Island (190)

1500   1600   1700   1800   1900   2000
-/-
26. Mar. 1800 discovered by Henry Waterhouse

Bollons Islands
210

North Cape   Anchorage
                      Bay
Windwards   Reef Point
Islands   NORTH
               PLAINS
Cave Point   Leeward
        Mount Galloway   Island
          366
Stack Bay
        Mount 361
        Waterhouse
                  Ringdove
                   Bay
                  Albatross
        South   Point
         Bay

0 1 2 3 4 5 km
|--|--|--|--|--|--|

Antipodes Island

EACH OF US LONGS for a doppelgänger who lives on the other side of the Earth, upside down, his feet facing ours, held on to the same globe by gravity. The people whose feet press against ours live on the same longitude but on opposing latitudes. Their seasons are the opposite of ours and their time is shunted forward. Our antipodes have summer when we have winter, and midnight when we have midday. But nobody lives on Antipodes Island – only a couple of fur seals and penguins with colourful crests. When Captain Henry Waterhouse discovers it on his way from Port Jackson to England, he calculates that the island is almost directly opposite the zero meridian of Greenwich. A reflection of a place, he thinks, a tiny doppelgänger of the British Isles. London, his birthplace, is as far from here as the North Pole is from the South, and so it does not matter which route home he takes. England and this place are points at either end of a skewer through the globe, an imaginary line through the centre of the Earth. // But the equation does not work. His homeland looks quite different.

This place here is mountainous, bare of trees and the climate is rough: cold and stormy. The mild air of the Gulf Stream is missing. Cattle that are brought here die quickly and quietly in the dun-coloured steppes of grass. And the thunderous echo of waves breaking against the hollows of the jagged coastline never ceases.

1° 18' S
90° 26' W

Floreana

Galapagos Islands (Ecuador)

Formerly *Charles Island*

SPANISH also known as *Santa María*

173 km² | 100 inhabitants

50 km
/→ Isabela Island

1050 km
--/--/--/--//→ Ecuador

830 km
--/--/--/-/→ Cocos Island (214)

Mar. 1535 discovered by Tomás de Berlanga

1500 /   *1600*   *1700*   *1800*   *1900*   *200*
-/-

1929 German settlement begins

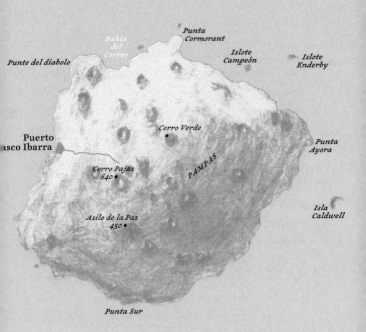

Punta
Cormorant

Bahía
del
Correo

Islote
Campeón

Islote
Enderby

Punte del díabolo

Cerro Verde

Puerto
asco Ibarra

Punta
Ayora

PAMPAS

Cerro Pajas
640 •

Isla
Caldwell

Asilo de la Paz
450 •

Punta Sur

0   1   2   3   4   5 km
|---|---|---|---|---|---|

Floreana

THE DRAMATIS PERSONAE: Dore Strauch, a teacher meant for greater things than a life lived by the side of a headmaster twice her age, and Dr Friedrich Ritter, a dentist from Berlin with a furrowed brow and glinting eyes, who wants to map the human brain, and feels that civilization has nothing more to offer him. In 1929, they leave their spouses to go to a place where the authority of the state ends and the law of necessity holds sway. // The scene: a lonely island which has resisted all attempts to colonize it. In the green crater of an extinct volcano, Friedrich and Dore establish their homestead Frido, build a hut of corrugated iron and cultivate an acre of land. // In their retreat from the world, they wear clothes only when there are visitors. At first it is the curious who come, those who want to supply the newspapers their stories of *Adam and Eve on Galapagos*. Soon after, the first imitators arrive. Ritter notes, *It is hard to believe that a patch of earth as difficult to reach as ours is visited so often*. // In 1932, a new settler steps on to the open-air stage: Eloise Wagner de Bousquet from

Austria, self-proclaimed baroness, a good-time girl with big teeth and thin eyebrows who wants to build a luxury hotel for millionaires. Her entourage includes cows, donkeys, chickens, eight hundredweight of cement and two lovers: Lorenz, a weedy young man with flaxen hair, and Philippson, a strapping young fellow; both slaves to her moods and desires. The baroness is soon playing the empress, tyrannizing the Ritters, ruling by whip and pistol, tormenting her lackey Lorenz and wounding animals only to nurse them back to health again. Her hotel remains unbuilt; the Hacienda Paradiso is a tent spread over four posts. // The comedy soon turns into a crime drama: in 1934, the baroness and Philippson vanish without a trace, Lorenz's skeleton is found on the beach of a neighbouring island and Dr Ritter dies of food poisoning after eating meat. Only Dore returns home to Berlin. The newspapers of the world speculate over the Galapagos Affair.

0° 51' S
169° 32' E

Banaba

(Kiribati)

Also known as *Ocean Island*

6.5 km² | 301 inhabitants

290 km
--/|→ Nauru

440 km
--/--/→ Gilbert Islands

1000 *1550 km*
--/--/--/--/--/--/|→ Howland Island (126)

20.*Aug.*1945 massacre of 143 Banabans by Japanese occupiers
1500 *1600* *1700* *1800* *1900* *200*
-/-

1804 discovered by Abraham Bristow

1900–79 phosphate minin

Tabwewa

86 •

Tabiang   Ooma

Lilian
Point   Home
         Bay   Sydney
                Point

0   1   2   3   4   5 km
/---/---/---/---/---/

Banaba

THE BANABANS' most important tool is made of wild almond wood and sharpened turtle shell. It is used to tattoo the skin with ink made of a dark paste of coconut ash mixed with salt and fresh water. The patterns are strictly prescribed. They are applied singly or repeated: straight and curved lines from which feathered lines grow. The head and the legs – practically the whole body – are tattooed. It is a preparation for the afterlife. // The souls of the dead journey to the west, where Nei Karamakuna, the woman with the bird's head, blocks the way and demands her favourite food: the patterns in their skin. She pecks the ink out of their limbs and their faces with her mighty beak, and grants spirit eyes to the dead in thanks, so that they can find their way to the spirit world. If the dead are not tattooed, Nei Karamakuna pecks their eyes out and they are condemned to wander in blindness for ever. // The Banabans do not bury their dead. They let the bodies hang from their huts until the flesh has decomposed, then wash the skeleton in the sea. The head is separated from the body;

the bones are stored under the houses and the skull is kept under the stone of the terraces, on which the young men play with the frigate birds. They dance around the tame birds and throw objects at them until they cannot move any longer and their wings are flattened to the ground. // Yet it was the birds who made this land. They nested on a swelling in the sea and left their droppings, which sank into the water and hardened into phosphate of lime on the reef. This layer grew many metres thick and rose above sea level, forming an island of pure phosphate.

Campbell Island
(New Zealand)

113.3 km² | uninhabited

660 km
--/--/-/⟶ New Zealand

1000   1900 km
--/--/--/--/-/--/--/-/⟶ Antarctica

730 km
--/--/--/⟶ Antipodes Island (154)

15.Oct.1995 meteorological station close▪

1500   1600   1700   1800   1900   ▪200▪
-/-
                              /
4.Jan.1810 discovered by Frederick Hasselborough

Isle de
Jeanette Marie

Borchgrevink
Bay

North Cape

Courrejolles
Point

Mount Faye
347

Gomez Island

Cook Point

•479
Mount Azimuth

Northeast Harbour

McDonald Point

Dent Island

Smoothwater
Bay

Penguin
Bay

Cattle
Bay

Complex
Point

Mount Lyall
420

East
Cape

ok Keys

Mount
Paris
•468

•Menhir

PERSEVERANCE HARBOUR

Erebus
Point

Monowai
Island

Rocky Bay

Mount Dumas
•500

Mount Honey
•558

South Point

sp Island

Survey Island

Antarctic
Bay

Shag Point

Southeast Harbour

Jacquemart Island

0   1   2   3   4   5 km
|---|---|---|---|---|

Campbell Island

ON 8 DECEMBER 1874 the sky clouds over; that night the weather is unsettled and it is misty. // There is a 60 per cent likelihood of being able to view the start of the transit of Venus here, and a 30 per cent chance of seeing the end: so Captain Jacquemart calculated when he spent nearly the whole of the previous December on the island. // Based on his findings, the Académie des Sciences decided to send an expedition here to view the transit. Sponsored by the government, the expensively equipped party leaves Marseille on 21 June, led by Anatole Bouquet de la Grye of the naval Hydrographic Office. // When Campbell Island finally appears out of the mist on 9 September, the men's first impression is of a sad place: a barren, treeless land with a plateau of straggly yellow bushes in the north and oddly shaped peaks in the south; the fjord of Perseverance Harbour in the middle. // On the morning of 9 December, the wind blows from the north-west, bringing scattered showers at about ten o'clock. The sky remains a solid grey until the warmth of the sun lightens the mist a little and its white disc finally appears behind the thick veil. Five minutes before Venus is to make its transit, the wind dies down. Bouquet de la Grye peeps through the eyepiece of the telescope at noon and cheers when he sees a dark patch at the edge of the sun: faint and jagged. It is Venus. Then a great cloud hides this rare event for more than a quarter of an hour. When it is gone, Venus is already covering half the sun. The outline of the planet is now quite distinct, entirely free of refractions of light or a halo. But this moment of clarity lasts no more than twenty seconds. // Then it is all over. Banks of fog roll in, making it impossible to see the sun again. When it clears hours later, Venus has long since disappeared into the sky.

Pingelap

Caroline Islands (Micronesia)

1.8 km² | 250 inhabitants

780 km
--/--/--// → Bikini Atoll

1000 *1990 km*
--/--/--/--/--/--/--/--/ → Papua New Guinea

1000 *1250 km*
--/--/--/--/--/-// → Banaba (162)

1792 discovered by Thomas Musgrave

1775 Typhoon Lienkieki wreaks devastation

1500 *1600* *1700* */1800* *1900* *2004*
-/--/--/--/--/--/--/--/--/--/--/--/--/--/--/--/--/--/--/-

1820s first instances of colour blindness

2000 the gene for achromatopsia is decoded

Takai

Tugulu

Pingelap

0   1   2   3   4   5 km
|---|---|---|---|---|

Pingelap

EVEN THE PIGS on this island are black and white. It is as if they have been made specially for the seventy-five people of Pingelap who see no colour: not the fiery crimson of the sunset, not the azure of the ocean, not the garish yellow of the ripe papaya, nor the ever-present deep green of the dense jungle of breadfruit palms, coconut palms and mangroves. // A tiny mutation in chromosome number eight and Typhoon Lienkieki, which laid waste to the island centuries ago, are responsible. Only twenty Pingelap inhabitants survived the typhoon and the famine that followed; one of them carried the recessive gene that soon made its presence known as a result of inbreeding. Today, 10 per cent of the population of Pingelap is completely colour blind, compared to a rate of one in thirty thousand elsewhere. // You can tell who they are by the way they lower their heads and blink constantly, by the way they flutter their eyelids and constantly screw their eyes shut, by the squint line above their noses. They avoid light, avoid the day, and often only leave their huts at dusk. The windows of their homes are covered in coloured film. // Many claim to always remember their dreams, and some say that they can see dark shoals of fish in deep water at night – they spot them by the faint moonlight reflected on the fins. // Their world may be grey, but they insist that they can see things hidden to those who see in colour: myriad shades and tones that cannot be imagined. Silly talk about the gloriousness of colour makes them indignant. Colour would only distract them from the essentials: the richness of shapes and shadows, forms and contrasts.

27° 9' S
109° 25' W

Easter Island

(Chile)

SPANISH *Isla da Pascua*
RAPANUI *Rapa Nui*, also known as *Te Pit Te Hunua*
['Navel of the World']
163.6 km² | 3,791 inhabitants

1000   2000   3000   **3690 km**
--/--/--/--/--/--/--/--/--/--/--/--/--/--/--/--/→ Chile

1000   2000   3000   **4190 km**
--/--/--/--/--/--/--/--/--/--/--/--/--/--/--/--/→ Tahiti

1000   2000   **2970 km**
--/--/--/--/--/--/--/--/--/--/--/→ Robinson Crusoe (122)

1687 sighted by the pirate Edward Davis

1500   1600   1700   1800   1900   200
--/--/--/--/--/--/--/--/--/--/--/--/--/--/--/--/--/--/--/-

9.Sept.1888 annexed by Chile

5.Apr.1722 (Easter Sunday) discovered by Jakob Roggeveen

Cabo Norte

Punta San Juan

*Caleta
Anakena*

Punta Rosalia

*BAHÍA DE
LA PÉROUSE*

Maunga
Terevaka
• 507

Maunga
Puakatike
370 •

• 302
Cerro Puhi

Volcán
Rano Raraku

P O I K E

Hatuiti

Maunga
O Tu'u
300 •

nta
ok

ga Roa

Cerro Tuutapu
270 •

Punta
Cuidado

Hanga Piko

eri

Maunga Orito
• 220

Punta
Baja

*RADA
BENEPU*

Punta
Redonda

Volcán
Rano Kao

Punta
Kikiri Roa

Cabo Sur

0 1 2 3 4 5 km
|---|---|---|---|---|

Easter Island

IT'S NO WONDER that Darwin never stopped here. Flora and fauna are scarce, and the abundance of the Galapagos Islands he was aiming for is weeks away by canoe. // No one knows now how high the giant palms that once covered the island grew. The sap that flowed from their trunks was fermented into wine sweet as honey; the wood was made into rafts and ropes to transport the statues. // Monsters dot the coastline. Hollow-eyed beings with elongated ears, weather-beaten skin and pouting lips like sullen children. These stone sentinels of hardened volcanic ash stand with their moss-covered backs to the sea, gazing into the palm forests with eyes of white coral on feast days. // The twelve tribes of Easter Island compete against each other: they make bigger and bigger monoliths, and secretly topple their rivals' statues in the night. They exploit and over-cultivate their pieces of earth, chop down the last tree, sawing off even the branch they are sitting on. It is the beginning of the end. They either die of smallpox, or become slaves on their own land, serfs working for the tenant farmers who turn their island into an enormous sheep farm. Out of 10,000, only 111 native islanders remain. Not a single palm tree is left, and not a single stone watchman remains standing. // Archaeologists later raise the monstrous figures and look for a trail. They search through piles of detritus, dig for seeds, collect bones and charred wood, try to decipher the sinuous etchings of Rongorongo glyphs, and wonder if anything in the stone countenances will tell them what has happened here. // Today, there is not a single tree on the barren land created from seventy volcanoes. But the airport's landing strip is so enormous that a space shuttle could touch down on it in an emergency. The end of the world is an accepted fact, and Easter Island is a case in point with its chain of unfortunate events that led to self-destruction; a lemming marooned in the calm of the ocean.

25° 3' S
130° 6' W

Pitcairn Island
(United Kingdom)

PITKERN *Pitkern Ailen*

4.5 km² | 48 inhabitants

480 km
--/--/→ Gambier Islands

1000   *2120 km*
--/--/--/--/--/--/--/-/→ Tahiti

1000   *2070 km*
--/--/--/--/--/--/--/-/→ Easter Island (174)

Jan.1790 settlement by the *Bounty* Mutineers
2.Jul.1767 discovered by Robert Pitcairn

1500   *1600*   *1700*   *1800*   *1900*   *200*
-/--/--/--/--/--/--/--/--/--/--/--/--/--/--/--/--/--/-

1856 resettlement on Norfolk Island

2002–5 rape trial

Western
Harbour

Adamstown

Point Christian 347•

Bounty Bay

Oh Dear

St Paul's Point

Down
Rope

Tautama

0   1   2   3   4   5 km

Pitcairn Island

THERE IS NO BETTER hiding place than this island, far off the trading routes and marked in the wrong position on admiralty charts. The sailors have mutinied and the afterlife has to be the judge of their actions. But there can be no homecoming – not for these men and the wives they have carried off from Tahiti. In England, they would be locked up; here in Pitcairn they are locked out. *Hiding here is just another way to die*, Fletcher Christian says, as they sit by the campfire. Two sailors later use its embers to set light to the *Bounty* by night in order to prevent any return to certain death on the gallows. Christian falls victim to a second mutiny, and more will follow. // *I want to find out what happened to the sailors after the mutiny. Why did they go to Pitcairn Island and within two years kill each other off? What is it in human nature that makes men violent even in an island paradise? That's what would interest me!* says Marlon Brando, who has artistic control written into every con-tract. // It is Christian's death scene: there he lies, just a head, blanket pulled up to his chin to hide his terrible burns. His face is damp with sweat and streaked with soot, and his eyes stretch wide, shining white in the darkness. The eyebrows lift and droop and Brando's quivering lips ask if he, Fletcher Christian, is dying. The man has only recently been a tinny-voiced posturing fop, all pomades and perfumes, a dandy of the South Seas slinking all over the 70-millimetre film in a silk dressing gown or lace jabot with a pink flower behind one ear, constantly dropping the clipped British accent he has practised. *What a useless way to die*, he says. His face freezes and the gaze is broken. The camera swings round and the burning *Bounty* sinks into the dark sea. The glittering curtains swish together and the most expensive film of all time comes to an end. But the island's story is far from its end.

51° 57′ N
179° 38′ E

Semisopochnoi

Rat Islands (United States)

RUSSIAN *Semisopochnoi* ['Has Seven Hills']
ALEUT *Unyax* or *Hawadax*

221.7 km² | uninhabited

--/--/--/--/--/--/₁₀₀₀ **1190 km**
—/→ Kamchatka

--/--/--/--/--/--/-/₁₀₀₀ **1360 km**
→ Cape Newenham

850 km
--/--/--/-/ → St George Island (202)

1741 discovered by Vitus Bering

1500   *1600*   *1700*   *1800*   *1900*   *200*
-/-

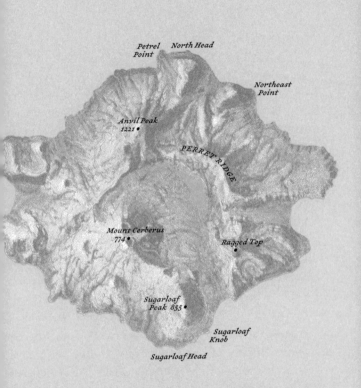

Petrel
Point

North Head

Northeast
Point

Anvil Peak
1221 •

PERRET RIDGE

Mount Cerberus
774 •

Ragged Top
•

Sugarloaf
Peak 855 •

Sugarloaf
Knob

Sugarloaf Head

0   1   2   3   4   5 km
|--|--|--|--|--|

Semisopochnoi

IT'S AN UNLIKELY partnership; a Russian name for an American territory. Semisopochnoi – perhaps the westernmost part of the United States. No one wants to know exactly. Nothing here is really important. No one has lived here – ever. There would be no reason to. Now and then, a few researchers come and collect rocks, measure the craters and take panoramic photographs in which the mountains look positively cinematic. A couple of arctic foxes take off into the undergrowth, staring at the visitors. They have no fear of these unknown creatures. The foxes' fur is a perfect deep blue. This land of seven hills is only one in a string of dancing pearls, a link that has broken free of the loose chain of islands that connects one continent to another – the proper hinterland that is later discovered to be the New World. // Here, on top of the Pacific ring of fire, the Earth mutters to itself, largely unnoticed by human beings. Volcanic eruptions are frequent but do not endanger anyone. Mount Cerberus is the liveliest of the volcanoes. Its three peaks watch over a rocky mountain terrain coloured crimson by the constantly overcast sky. A few of the craters emit small puffs of smoke from time to time, which could just as easily be the clouds that hang from the peaks.

10° 18′ N
109° 13′ W

Clipperton Atoll

(France)

FRENCH *Île Clipperton* or *Île de la Passion* ['Passion Island']

1.7 km² | uninhabited

1080 km
--/--/--/--/-/⟶ Mexico

1000   2000 *2260 km*
--/--/--/--/--/--/--/--/--/--// ⟶ Galapagos Islands

950 km
--/--/--/--/--/ ⟶ Socorro (194)

1892–7 guano extraction by the *Oceanic Phosphate Company*

1500   1600   1700   1800   /1900   2000
-/--/--/--/--/--/--/--/--/--/--//--/--/--/--/--/--/--/--/--/--/--/--/--/--/--/--/-

3.Apr.1711 *(Good Friday)* discovered by Martin de Chassiron
and Michel du Bocage

Baie
de la
Pince

Grand
Récif

Le Rocher

0   1   2   3   4   5 km
|---|---|---|---|---|

Clipperton Atoll

THE SHIP FROM ACAPULCO does not arrive. An American cruiser brings the news: the world is at war and Mexico is in chaos. They have forgotten about them. Their general is no longer in power. // Not a blade of grass grows on the island. A dozen scraggy pigs sprawl under the palm trees, descendants of a stranded herd. They eat the orange land crabs, of which there are millions on the island. It is not possible to take a step here without treading on a shell. There is a crunching sound when the governor, Capitán Ramón de Arnaud, walks over the island. As always, he is attired in Austrian parade ground uniform, and his wife in an elegant evening gown, with diamonds on her fingers and round her neck. That day, he announces that *Evacuation is not necessary. Orders are orders*. The garrison stays put: fourteen men, six women and six children. No ship arrives: not the one from Acapulco or any other. Supplies run low. There is an outbreak of scurvy: gums bleed, wounds fester, muscles waste away, limbs rot and there is heart failure. They dig deep graves for the dead to protect them from the greed of the crabs. // At some point, the governor can stand the screeching of the seabirds and the roar of the sea no longer. He thinks he sights a ship, and sets off in a small boat. The remaining soldiers drown along with him. Now there is only one man left on the island: Victoriano Álvarez, the former keeper of the lighthouse whose light no longer burns. He proclaims himself king, the king of Clipperton, he takes mistresses, he rapes and he kills. He rules for almost two years. // On 17 July 1917, the women strike him down with a hammer and smash his face in. A ship appears on the horizon. The women and children wave at it while the crabs scuttle towards the lighthouse, lured by the approach of fresh blood. A boat lands on the Phosphate Company's old quay and the four surviving women and their young children leave the loneliest atoll in the world. Looking back from the USS *Yorkmen*, they can see the orange ring of crabs in the lagoon for a long time.

Raoul Island

Kermadec Islands (New Zealand)

Formerly *Sunday Island*

29.4 km² | 10 residents

910 km
--/--/--/-/ ⟶ Tonga Islands

980 km
--/--/--/--/ ⟶ New Zealand

1000 *1370 km*
--/--/--/--/--/-/ ⟶ Norfolk Island (146)

21. Nov. 1964 volcanic erupti

1500   *1600*   *1700*   *1800*   *1900*   *200*
-/--/

18. Mar. 1793 discovered by Joseph Bruny d'Entrecasteaux

1937 nature reserve is opened

Napier Island

Hutchinson
Bluff

Meyer Islands

Meteorological
Station

Herald Islets

•455
Pukekohu

DENHAM
BAY

Moumoukai
Prospect   •516
•498

Lava Point

Wilson Point

Nash Point

Smith Bluff

D'Arcy Point

0   1   2   3   4   5 km

Raoul Island

EVERY YEAR, the New Zealand Department of Con servation (DOC) sends staff to live for twelve months on the otherwise uninhabited island. Nine volunteers assist them over the summer or the winter months and they stay for up to six months. But the DOC warns in its information brochure: *It takes a special person to cope with living on an island as isolated as Raoul. He or she must be able to carry out a wide range of practical tasks – from weed pulling to track maintenance, fixing buildings to baking bread. // The Raoul volunteer pro gramme provides the opportunity to experience a remote offshore island and explore an unusual subtropical eco system. There are many challenges to living and work ing on Raoul. The area is actively volcanic and earth quakes are a regular fact of life. The terrain is very steep and rugged. The duties are often repetitive and exhausting. Much of the work involves weeding alien plants. // Once there, you are there for the full term of the volunteer stint. Mail is sometimes dropped by pass ing RNZ Air Force planes or private boats. The fastest emergency air rescue is at least 24 hours away. // Raoul volunteers must be adaptable, cautiously adventurous, content to amuse themselves and also happy to work in a small team. // Applicants will need to be physically fit and agile, experienced at moving through bush without a track. Climbing skills and practical experience in maintaining buildings and facilities are an advantage. Applications can be sent to: Department of Conservation, PO Box 474, Warkworth, New Zealand.*

Socorro Island

Revillagigedo Islands (Mexico)

SPANISH also known as *Isla Santo Tomás*,
formerly *Isla Anublada* ['Cloudy Island']

132.1 km² | 250 residents

50 km
/⟶ San Benedicto

460 km
--/--/⟶ Lower California

1000　　　　2000　　　　3000　　　4000*8460 km*
--/--/--/--/--/--/--/--/--/--/--/--/--/--/--/--/···/⟶ Taongi (1.

21. Dec. 1533 discovered by Hernando de Grijalva

1957 military base constructed

1500　　　1600　　　1700　　　1800　　　1900　　 200
-/--/--/--/--/--/--/--/--/--/--/--/--/--/--/--/--/--/-

Early 1920s George Hugh Banning visits

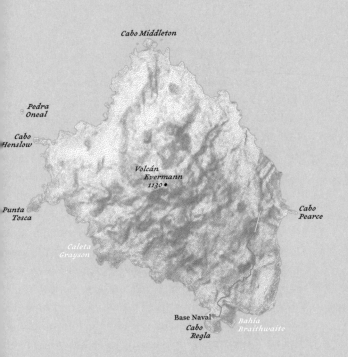

Cabo Middleton

Pedra
Oneal

Cabo
Henslow

Volcán
Evermann
1130 •

Punta
Tosca

Cabo
Pearce

Caleta
Grayson

Base Naval
Cabo
Regla

Bahía
Braithwaite

0   1   2   3   4   5 km
|--|--|--|--|--|--|--|--|--|--|

Socorro Island

AS THEY ENTER Braithwaite Bay, the island looks like a house that has been locked fast. The water appears dead and the wet shingle beach shines coldly under cliffs of lava and hills covered with prickly shrubs. One sailor makes a brief excursion to the island that evening and returns downcast, as if he has seen something parcticularly dispiriting. // The next day, George Hugh Banning, second mate on the *Velero II*, sets out alone at the crack of dawn to explore the scorched island. He first discovers sheep on a plateau – panicked by the intruder, they thunder down the slopes and disappear into the undergrowth – then finds the wild descendants of a small herd of pigs set down on the island by whalers. No one knows where they find their water, as according to the American navy there is no fresh water on Socorro Island. Banning follows the pigs, making his way through the scrubland: a labyrinth of thorny creepers several metres high, ragged tree stumps and shrivelled vines. With every step there is a snap and crunch, with every snap a scratch, with every crunch a blow; every time he trips, cactus thorns dig into his ankles, calves and hands. Over and over again he crawls through the unyielding brushwood on his hands and knees and climbs over thorny branches of the prickly pear. Soon he is in the deepest bush, so dense that even the sheep cannot get through. Banning looks around him: this is not a wood any longer, it is a wilderness. The thick canopy of leaves does not let any light through. Gigantic snakes seem to be slithering along the branches and every bare tree looks like a creature being tortured; the crowd of bony shapes seems to be closing in on him from all sides. This is what hell must look like. Banning, blundering on blindly, begins to feel as if soon he might even bump into himself. Filled with fear and driven by desperation, he reaches for his dagger and starts running. He tramples everything in his path as he fights his way out of the virgin bush. Finally he is in the open again – breathless and covered in scratches.

Iwo Jima

Volcano Islands (Japan)

Also known as *Sulphur Island*

JAPANESE *Iōtō* ['Sulphur Island']

23.2 km² | 370 residents

1050 km
--/--/--/--//→ Tokyo

1000   1950 km
--/--/--/--/--/--/--/--/→ Taiwan

1000   2000   3140 km
--/--/--/--/--/--/--/--/--/--/--/-/→ Atlasov Island (138)

19 . Feb. – 26 . Mar. 1945 Battle of Iwo Jima

1500   1600   1700   1800   1900   200.
-/-

1968 returned to Japan

Kitano-hana

Hiraiwa-
wan

Kangoku-iwa

MOTO-
YAMA

169 •
Suribachi-yama
Tobiishi-hana

0 1 2 3 4 5 km
|---|---|---|---|---|

Iwo Jima

THE HORIZON IS aslant and the sky overcast with clouds and plumes of smoke from the exploding mines and bombs. At the summit of Mount Suribachi, six soldiers ram a flagpole into the rubble on the ground. Using all their strength, they raise it above their heads. Faceless figures helping each other: one kneeling on the rocks, another reaching upwards. Joe Rosenthal presses the shutter release button: this moment, lasting four-hundredths of a second, on 23 February 1945, becomes the most famous war photograph of all time. // The men give their lives for the flag of their distant country. A gesture of courage; a forlorn act of representation: stars and stripes, red, white and blue, hands on hearts. The Americans occupy the highest peak on Iwo Jima: a 169-metre-high barren mound of ash on the southern tip of the tiny island that has suddenly acquired strategic importance. It is an unsinkable aircraft carrier near the enemy country, large enough for bombers to take off from and land on in the future. // It is a photograph of a prematurely declared victory, for the island is not yet conquered, the battle not yet won. The enemy lurks, hidden in the volcanic landscape, tossing hand grenades out of underground chambers. The labyrinth of 1,000 man-made caves turns into a graveyard for 20,000 Japanese soldiers. // The roll of film is flown to Guam and developed in the headquarters of the Wartime Still Picture Pools. Before the day is out, the iconic shot is discovered: a photograph akin to a statue, trimmed to portrait size. A teleprinter transmits it to the homeland and it immediately makes the front page of all the Sunday papers. A few months later it appears on a stamp; ten years later it is transformed into the largest bronze monument in the world – soldiers twenty-four metres high on a granite base – at a military cemetery near Washington. // This image is now part of every battle scene. Three New York firemen raise the flag in the dusty ruins of one September – the summit of Suribachi is now called Ground Zero.

56° 35' N
169° 36' W

St George Island

Pribilof Islands (United States)

90 km² | 128 inhabitants

1000 *1240 km*
--/--/--/--/--/→ Anchorage

1000 *1630 km*
--/--/--/--/--/--/-/→ Kamchatka

1000   2000   3000   *4340 km*
--/--/--/--/--/--/--/--/--/--/--/--/--/--/···/→ Lonely Island

25.Jun.1786 discovered by Gavriil Pribilof
1500   *1600*   *1700*   *1800*   *1900*   *2000*
-/-
1786 extinction of Steller's sea cow

Suskaralogh
Point

St George

High Bluffs
MAYNARD
HILL

First Bluffs

Tolstoi
Point

Rush
Point

•309

202•

Sea Lion
Point

ZAPADNI
BAY

ULAKAIA
HILL

Garden
Cove

SOUTH
HILL

Cascade
Point

0   1   2   3   4   5 km

St George Island

ITS SHAPE IS BOTH STRANGE and marvellous. The Arctic sea cow – seen alive only by Georg Wilhelm Steller and the sailors who hunted it to extinction – must have lived on the shores of this island in the furthest reaches of the ocean. All that remains of the species is a few skeletons, a couple of scraps of skin and Steller's description, recorded when he was shipwrecked on Vitus Bering's second expedition to Kamchatka. The sea cow belongs to the sirenian order and it does, indeed, have the forked tail and breasts of a mermaid. Its skin is several centimetres thick and feels like the bark of an ancient oak, its back is black, smooth and hairless and its neck is wrinkled. Its forelimbs are two stunted flippers. Its head is like no other animal's: small and squarish, with no neck, on top of a monstrous body. Its nostrils are like those of a horse and its ears are little more than two tiny holes. Its eyes have no lids, and are no bigger than a sheep's: the iris is black and the pupil yellowish-blue. // The insatiable sea cow grazes continually in the shallows, its huge body only half-submerged in water as it crushes sea grass in its toothless jaws. Seagulls perch on its back, picking vermin off it. It comes up for air every four to five minutes, breathing noisily. // These sea creatures mate on quiet spring evenings when the weather is calm: *in the way that humans do. The male lies on top and the female below*, Steller records. And they embrace each other. They often come so close to land that human beings can gently pat them; they can also kill them. The sea cow is mute; but when it is wounded, it heaves a deep sigh.

Tikopia *Santa Cruz Islands*
(Solomonen Islands)

4.7 km² | 1,200 inhabitants

210 km
--/--/→ Vanikoro

1100 km
--/--/--/--/-/→ Fiji

1000   1540 km
--/--/--/--/--/--// → Takuu (218)

1928/29 Raymond Firth's first held expedition

1500   1600   1700   1800   1900   200
-/-

1606 discovered by Pedro Fernández de Quirós

Dec. 2002 devastated by Cyclone Zoe

RAVENGA

Fatapu Point

380 •

Reani

Rakionamo Point

Tereufa Point

Fono vai
Korokoro Point

Sautafi

FAEA

Lake
Te Roto

Matautu

Atunu

Asanga

Ratea

0   1   2   3   4   5 km
|---|---|---|---|---|

Tikopia

HUMAN BEINGS HAVE LIVED on this island for three thousand years. It is so small that the breaking waves can be heard even from the centre of the island. The Tikopians catch fish in the brackish water of their lake and shellfish from the sea. They cultivate yams, bananas and giant swamp taro, and bury breadfruit for lean times. That is enough to keep 1,200 people but no more. // When a cyclone or drought destroys the crops, many islanders opt for a swift death. Unmarried women hang themselves or swim out into the open sea, and some men go out to sea with their sons in a canoe, never to return. // Every year, the chiefs of the four clans preach the ideal of zero population growth. All the children in each family must be able to live from the land it owns, so only eldest sons can start families. The younger sons stay single and are careful not to produce any children. Feeling obliged to hinder conception, the men practise coitus interruptus, and if this does not work, the women press hot stones to their pregnant bellies. // A couple stop having children when the eldest son is old enough to marry. This is when a man will ask his wife, *Whose child is this, for whom I must fetch food from the field?* He decides whether the baby lives: *The plantations are small. Let us kill the child, for if it lives, it will have no garden.* The newborn is laid on its face to suffocate. There are no funerals for these children: they have not participated in life on Tikopia.

18° 7' N
145° 46' E

Pagan

Mariana (United States)

SPANISH formerly *San Ignacio*

47.2 km² | uninhabited

310 km
--/-/--→ Saipan

1000   *2000*   *2670 km*
--/--/--/--/--/--/--/--/--/--/--/-/--→ Manila

840 km
--/--/--/-/--→ Iwo Jima (198)

1669 discovered by Diego Luis de Sanvitores

1500   *1600*   *1700*   *1800*   *1900*   *2000*
-/-/-

1981 evacuated after volcanic eruption

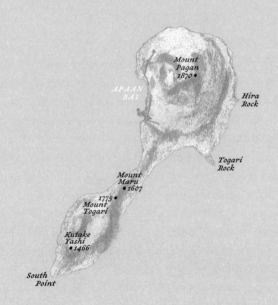

Mount
Pagan
1870 •

*APAAN
BAY*

*Hira
Rock*

*Togari
Rock*

Mount
Maru
• 1607

1775 •
Mount
Togari

Kutake
Yashi
• 1466

*South
Point*

0   1   2   3   4 • 5 km
|---|---|---|---|---|

Pagan

THE TALLEST MOUNTAIN range in the world is underwater – where the Pacific plate converges with the Philippine plate in the Marianas Trench, several kilometres deep – and its smoking volcano cones rise out of the ocean. // Pagan is a double island of two of these volcanoes held together by a land mass. At its narrowest point, it is only a few hundred metres wide. // The village of Shomushon lies at the foot of Mount Pagan in the north. Its people want to be evacuated because smoke has been rising from the summit for some time, and there have been earthquakes. But no one takes any notice. They say the volcano is not dangerous. // On 15 May 1981, it erupts, spewing fire, hurling rocks and shooting fountains of lava into the air. The sky turns black; it rains ash and smells of sulphur and burning earth. The raised huts in Shomushon shake and a flood of lava spreads through the palm trees. Soon the first crackle of fire in the village is heard. The mayor sends a message by short-wave radio – *This is it! Come get us!* – before the sixty villagers flee, crossing the narrow neck of land to the south.

They take refuge behind a mountain ridge and pray to be spared from the glowing river. // When they are evacuated by air shortly after, only the rooftops of Shomushon can be seen above the layer of brown ash. On Pagan, there are now 20 million tonnes of tuff stone, the material of the Colosseum, the Pantheon and the Baths of Caracalla.

Cocos Island

(Costa Rica)

SPANISH *Isla del Coco*

24 km² | uninhabited

550 km
--/--//→ Puntarenas

1000 km
--/--/--/--/→ Colombia

1000   *2000*   *2500 km*
--/--/--/--/--/--/--/--/--/--/→ Clipperton Atoll (186)

1526 discovered by Juan Cabezas

1500/   *1600*   *1700*   *1800*   *1900*   *2000*
-/-

11. Nov. 1897 August Gissler becomes governor of the island

Isla
Manuelita

Bahía
Chatham

Isla Cónico

Bahía
Wafer

Punta Gissler

Punta María

Cerro
Iglesias
634

Río Genio

Cabo Atrevido

Cabo Lionel

Cabo Descubierta

Islas Dos Amigos

Bahía Iglesias

Bayo
Alcyone

Cabo Dampier

0 1 2 3 4 5 km
|--|--|--|--|--|--|

Cocos Island

ONE ISLAND, TWO MAPS, three treasures. August Gissler is certain that he can find the pirate gold of the black-flagged ships that have rounded Cape Horn: Edward Davis's booty, Benito Bonito's loot and the cathedral treasure of Lima, with its life-sized statue of the Blessed Virgin in solid gold. // The son of a manufacturer from Remschied, Gissler prefers a sailor's life to managing a paper factory. He looks at the crosses on the map and reads the descriptions: *in the north-eastern corner of Wafer Bay, in a small grotto at the foot of a cliff with a triple jagged edge, 200 feet under the flood line*. He is thirty-two years old – a tall man with light eyes and a full beard. When he strikes his spade into the ground at this spot, he finds nothing more than wet earth. Gissler digs one hole after another, holes so deep that he stands in groundwater up to his ankles, so wide that ships, though not dreams, could be buried within them. // He buys more maps in harbour gin joints – maps that pirates' grandsons have inherited, with crosses old and new – and digs fresh holes in dark loam. With his pickaxe and his spade, he digs in circles, and issues new shares in his homeland for his Cocos Plantation Company, shares in the island of gold. Six German families and Gissler's wife follow him there. They settle in the bays of the tropical island; they build log cabins and cultivate coffee, tobacco and sugar cane; they dig and dig and find nothing. // Three years later, the Gisslers are on their own again, masters of the hidden riches. To search is more blessed than to find, Gissler thinks. Every empty hole is only proof that the treasure must be somewhere else on this 2,400-hectare piece of land. // When he finally leaves the island, in 1905, he has dug so many holes that his beard reaches to his knees. He has lost sixteen years of his life. All he has found are thirty gold pieces and one golden glove. Shortly before his death in New York on 8 August 1935, he is still saying, *I'm sure there is great treasure in the island. But it will take a lot of time and money to unearth it. If I were young, I would start again from the beginning.*

4° 45' S
156° 59' E

Takuu

(Papua New Guinea)

Also known as *Mortlock Islands*

ÉTAKUU also known as *Tauu*

1.4 km² | 560 inhabitants

220 km
--/→ Bougainville

510 km
--/--/→ New Britain

1000 1280 km
--/--/--/--/--/→ Pingelap (170)

1500   1600   1700   1800   1900   200
-/-
                              /
19. Nov. 1795 sighted by James Mortlock

Nukerekia

TAAKAU PASSAGE

LAGOON

Lotuma Maturi
Farefatu

Nukutuurua
Karuteke
Nukuaafare

Nukutoa Petasi

Takuu

AVA PASSAGE

0 1 2 3 4 5 km

Takuu

NEITHER MISSIONARIES nor researchers are allowed on the island. The people of Takuu wish to stay true to themselves and to their beliefs. They need the presence of the spirits who built this atoll out of the bones of the ocean, and of their ancestors, who have protected it ever since, this brittle ring of sand only one metre above the high tidemark. // The sea level is rising and the wind is blowing. This island is sinking. The beach is narrower after every storm. Entire pieces of land disappear overnight. The movement of the tectonic plates and the changing climate are responsible. The sea is gobbling up more and more of the land. It is now covering the roots of the coconut trees and turning the groundwater brackish, so the taro plants are withering and meals are too meagre to stave off hunger. // The old people do not believe the island is sinking. They refuse to leave it. They build dykes instead, packing rocks and brushwood into wide-meshed nets and tossing them on to the shores that have been diminished; they call on the spirits and their ancestors for help. // The young people do not think at all – neither about the future nor about the past. All day long, they drink the juice of the coconut palm, fermented in the hot sunlight. Plastic bottles gather round the tops of the trees to collect more. // Takuu will sink – next month, next year.

ANTARCTIC
OCEAN

Laurie Island

Deception Island

Peter I Island

Franklin Island

60° 44' S
44° 31' W

Laurie Island

South Orkney Islands (Antarctica)

SPANISH *Lauría*

86 km² | 14–45 residents

810 km
--/--/--//→ South Georgia

1000 1280 km
--/--/--/--/--/--//→ Falkland Islands

250 km
--/→ Deception Island (228)

6.*Dec.*1821 discovered by George Powell and Nathaniel Palmer

1500   1600   1700   1800   1900   200
-/-

21.*Mar.* – 26.*Nov*1903 the Scottish National
Antarctic Expedition winter on Laurie Islan

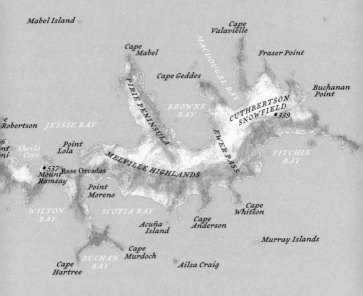

Mabel Island

Cape
Valavielle

Cape
Mabel

Fraser Point

MACDOUGAL BAY

PIRIE PENINSULA

Cape Geddes

BROWNS
BAY

CUTHBERTSON
SNOWFIELD

Buchanan
Point

• 339

Robertson

JESSIE BAY

EWER PASS

FITCHIE
BAY

Point
Lola

Sheila
Cove

MELVILLE HIGHLANDS

• 537
Mount
Ramsay

Base Orcadas

Point
Moreno

Cape
Whitson

WILTON
BAY

SCOTIA BAY

Cape
Anderson

Acuña
Island

Murray Islands

Cape
Murdoch

BUCHAN
BAY

Ailsa Craig

Cape
Hartree

0   1   2   3   4   5 km
|--|--|--|--|--|

Laurie Island

ALLAN GEORGE RAMSAY is dying. On the voyage from Troon to the Cape Verde Islands, he feels an odd piercing sensation in his chest, and when they stop on the Falkland Islands for a few weeks, these insidious attacks only become more frequent. In the end, he can deny it no longer: he, the chief engineer on the *Scotia*, is seriously ill. But Ramsay decides not to tell anyone. What could he do? Report his condition to the expedition leader and ask to be sent back to Scotland, knowing that they would never find a replacement for him, here in the middle of nowhere? He has no choice. Besides, he wants to see with his own eyes the white walls of the south, the jutting mountains of ice: the Antarctic land. // He sees it in February, when they could press southwards no more, and decide to winter on this island. When they finally find a safe bay to anchor in after searching for several days, Ramsay is no longer able to work. While the other crew members cover the *Scotia* in a canopy of snow, build two huts, take notes on the penguin colonies, conduct meteorological readings and magnetic observations,

Ramsay spends most of the time wrapped in woollen blankets by the stove in the ship's cabin. On 6 August 1903, he dies of heart failure. Two days later, they bury him on the rocky north beach of Scotia Bay, in the shadow of a mountain that they name after him. All the members of the Scottish National Antarctic Expedition and a few Adelie penguins pay their respects. Wearing a kilt, the laboratory assistant Kerr plays the Scottish lament on the bagpipes: *I've heard them lilting, at the yowe milking / Lassies a-lilting before dawn of day / But now they are moaning on ilka green loaning; / 'The Flowers of the Forest are a' wede away' / The Flooers o' the Forest, that fought aye the foremost / The pride o' oor land lie cauld in the clay.*

62° 57' S
60° 38' W # Deception Island

South Shetland Islands (Antarctica)

SPANISH *Isla Decepción*

98.5 km² | uninhabited

20 km
/→ Livingston Island

100 km
-/→ Antarctic Peninsula

1000   1490 km
--/--/--/--/--/--/→ Peter I Island (236)

1967–70 volcanic eruptions

15 . Nov . 1820 Nathaniel Palmer discovers the entry into the caldera

1500   1600   1700   1800   1900   2000
-/-

1906–31 whaling operation

29 . Jan . 1820 probable sighting by Edward Bransfield and William Smith

Macaroni Point

Goddard Hill
• 332

KENDAL TERRACE

Telefon
Bay
Pendulum
Cove

Mount Pond
• 539

STONETHROW RIDGE

PORT FOSTER

Fumarole
Bay

Baily Head

Base
Decepción

Sewingmachine
Needles

Gabriel de Castilla
Station

NEPTUNES
Neptunes
Window

BELLOWS

Mount Kirkwood
• 452

Entrance
Point

New Rock

South Point
Låvebrua
Island

0   1   2   3   4   5 km
|---|---|---|---|---|---|

Deception Island

THE ENTRANCE to the caldera is easy to miss: it is less than two hundred metres wide. Here in Neptunes Bellows, at the gates of hell, in the jaws of the dragon, the waves crash interminably. Behind it, hidden beneath the dozing volcano, is one of the safest harbours in the world: Whalers Bay. The inhabitants call this place New Sandefjord. It is the southernmost whale oil processing plant in the world, with its own fleet: two triple-masted ships, eight small whaling steamers and two large ones. Apart from a handful of Chilean stokers, two hundred Norwegians live here, along with one woman: Marie Betsy Rasmussen, the first female ever to be in Antarctica. She is the wife of Captain Adolf Amandus Andresen, the manager of one of the three companies who have been whaling here for two years. // The season runs from the end of November to the end of February. They hunt whales with new techniques tested in the north. Harpoons shoot out of the guns on the foredeck with explosive force, burying deep into the backs of the large animals, which all the whalers can identify from a distance. The humpback whale spouts a low plume of water and has a bump on its back. They can recognize the fin whale from the height of its spray. They identify the blue whale, the most precious of all, by its dorsal fin and the high column of water it spouts. A steamer captures up to six whales and ships them to the bay in the evening. On the dark beach, the whalers hack the baleen away from the jaws, pull the skin off, separate blubber from flesh, and boil the white gold in giant containers to extract the whale oil. The vessels are not heated with coal, but by burning the bodies of penguins they have caught from Baily Head. // They leave the rest to rot. The whale skeletons show white against the dark sand, the water is red with blood and the stench of rotting flesh fills the air. Thousands of plundered bodies decompose in the crater's overflowing pond.

Franklin Island

(Antarctica)

33 km² | uninhabited

70 km
–/⟶ Victoria Land

150 km
–/⟶ Ross Island

1000   *2000* *2410 km*
––/––/––/––/––/––/––/––/––/–/⟶ Macquarie Island (130)

27. Jan. 1841 discovered by James Clark Ross

1500   *1600*   *1700*   *1800*   *1900*   *2000*
–/–

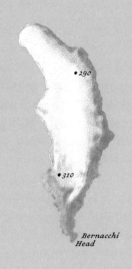

290

310

Bernacchi Head

0 1 2 3 4 5 km
|---|---|---|---|---|

Franklin Island

HMS *Terror* AND HMS *Erebus* have proved themselves in the ice. The bomb vessels look unwieldy, like shoeboxes, but they have robust hulls and carry fifteen tonnes of heavy steam engine in each armoured belly. They are warships, refitted for battle against the ice. One morning, when the fog finally lifts, they find themselves lying in a deep, white bay that stretches all the way to an island. Accompanied by a few officers, Captain James Clark Ross leaves the *Erebus* and rows towards the island, followed by Commander Francis Crozier with another group from the *Terror*. The waves are high and choppy, so the captain disembarks from a whaleboat and jumps on to a rock in one daring leap. The others follow him by rope. It is so cold that the rocks are covered in a film of ice. // This island is nothing but volcanic rock. The black cliffs in the north have flawless white stripes a few feet wide running through them. There is not the slightest sign of vegetation anywhere. To everyone's satisfaction, Captain Ross names the island in honour of His Excellency Sir John Franklin of the Royal Navy, hero of

Trafalgar, lieutenant-governor of Van Diemen's Land and polar explorer, who still dreams of finding the Northwest Passage. // Four years later, Franklin will set off in search of the short cut through the ice, the fabled passage to the Orient. Only two ships are suitable for this expedition: the Terror and the Erebus. Francis Crozier, first officer but forever second, is given command of the Terror. The ships become icebound near the north coast of King William Island. // James Clark Ross is among those who sets off with ships and sled dogs to find them. He will not find Sir John, his friend Crozier or the two bomb vessels, *Erebus* and *Terror*, with which he had charted the Antarctic coast. The horror and darkness of their fate remain hidden. A small island of volcanic rock is Franklin's memorial, but his grave lies under the ice at the opposite pole.

68° 53' S
90° 34' W
Peter I Island

(Antarctica)

RUSSIAN *Ostrov Petra I*
NORWEGIAN *Peter I Øy*

156 km² | uninhabited

420 km
--/-/⟶ Antarctica

1000   *1850 km*
--/--/--/--/--/--/--/-/⟶ Cape Horn

1000   *2000*   *3040 km*
--/--/--/--/-/--/--/--/--/--/--/--//⟶ Franklin Island (232)

2. Feb. 1929 landed on by Ola Olstad

1500   *1600*   *1700*   *1800*   *1900*   *2000*
-/-

21. Jan. 1821 discovered by Fabian Gottlieb von Bellingshausen

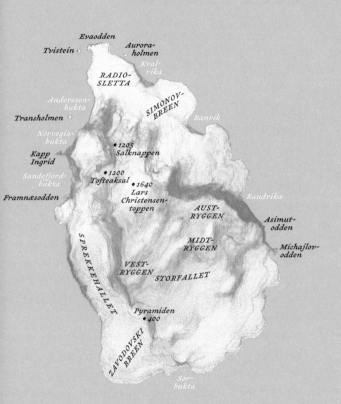

Evaodden

Tvistein

Auroraholmen

Kvalvika

RADIOSLETTA

SIMONOVBREEN

Ranvik

Anderssenbukta

Transholmen

Norvegiabukta

Kapp Ingrid

• 1205
Salknappen

Sandefjordbukta

• 1200
Tofteaksal

• 1640
Lars Christensentoppen

Framnæsodden

AUSTRYGGEN

Raudvika

Asimutodden

Michajlovodden

MIDTRYGGEN

S P R E K K E H A L L E T

VESTRYGGEN

STORFALLET

Pyramiden
• 400

ZAVODOVSKI
BREEN

Sørbukta

0   1   2   3   4   5 km
|---|---|---|---|---|

Peter I Island

LARS CHRISTENSEN, shipowner and consul of Sande-fjord, is equipping his whaling ship SS *Odd I* for an expedition. Laden with coal, the whaling vessel leaves the harbour of Deception Island on 12 January 1927. Five days later, it reaches Peter I Island. Discovered over a century earlier, the island is covered in pack ice almost year round, and no one has yet landed on it. They sail round it. The island's highest peak is a volcano on the west coast; no one knows if it is merely asleep or has been silenced for ever. The shore is bare and steep; cliffs of ice rise almost vertically out of the stormy sea. That afternoon, the captain attempts to land, but in vain. The island does not have a single sheltering bay, let alone a harbour; only a couple of narrow beaches of black rock and glaciers that stretch their tongues into the ocean. It is impossible to reach. In order to take something with them, they collect scattered rocks as proof of their journey. // The petrographer Olaf Anton Broch examines them thoroughly: *The specimens at hand, 175 pieces in total, are mostly rounded pebbles from the beach, varying between the size of a hazelnut and two balled fists. Some of them have a less dense, almost clinker-like consistency. They were collected in the west coast area, off Cape Ingrid Christensen. The types of rock are all represented by several specimens and are more or less related to each other petrographically. All the specimens examined are of an eruptive nature. On a macroscopic level, one has the impression that there is a great variety of different types of rock, but a more detailed examination reduces the number of categories. An analysis of 22 thin sections shows that the samples consist solely of basalt, andesite and what is called trachyandesite. Basalt pebbles were in the significant majority. Peter I Island is predominantly basalt in character.* There is nothing more to say about a land that no one has set foot on.

PENGUIN BOOKS

Published by the Penguin Group
Penguin Books Ltd, 80 Strand, London WC2R ORL, England
Penguin Group (USA) Inc., 375 Hudson Street, New York, New York 10014, USA
Penguin Group (Canada), 90 Eglinton Avenue East, Suite 700,
Toronto, Ontario, Canada M4P 2Y3 (a division of Pearson Penguin Canada Inc.)
Penguin Ireland, 25 St Stephen's Green, Dublin 2, Ireland (a division of Penguin Books Lt
Penguin Group (Australia), 707 Collins Street, Melbourne, VIC 3008, Australia
(a division of Pearson Australia Group Pty Ltd)
Penguin Books India Pvt Ltd, 11 Community Centre,
Panchsheel Park, New Delhi – 110 017, India
Penguin Group (NZ), 67 Apollo Drive, Rosedale, North Shore 0632, New Zealand
(a division of Pearson New Zealand Ltd)
Penguin Books (South Africa) (Pty) Ltd, Block D, Rosebank Office Park, 181 Jan Smuts
Avenue, Parktown North, Gauteng 2193, South Africa

Penguin Books Ltd, Registered Offices: 80 Strand, London WC2R ORL, England

www.penguin.com

First published in German as *Atlas der abgelegenen Inseln* 2009
This translation first published by Particular Books 2010
Published in this format 2012
011

Copyright © mareverlag, Hamburg, 2009, 2012
Translation copyright © Christine Lo, 2010, 2012

The moral right of the author and the translator has been asserted

ORIGINAL ILLUSTRATIONS AND DESIGN Judith Schalansky, Berlin
FONT MVB Sirenne
PAPER Oria offset Neutro fsc
PRINTING AND BINDING Printed in Italy by Graphicom srl

All the islands in this atlas are depicted on a scale of 1:200000

A CIP catalogue record for this book is available from the British Library

978-1-846-14349-6

www.greenpenguin.co.uk

Lonely Island

Semisopo

Atlasov
Island

Iwo Jima

Pagan

Taongi

Pingelap

Howland Isla

Diego Garcia

Bana

Takuu

South Keeling
Island

Christmas Island

Tiko

Norfolk Island

Amsterdam Island

Saint Paul Island

Campbell Islan

Macquarie Island

Franklin Islan

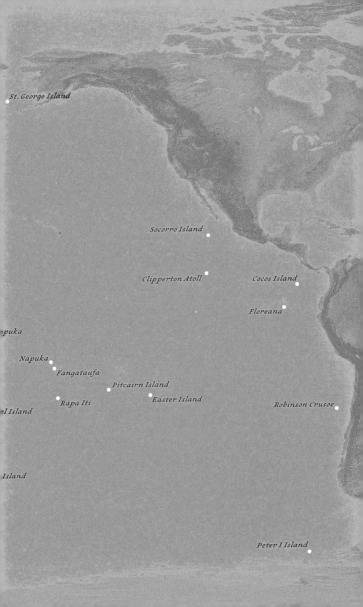

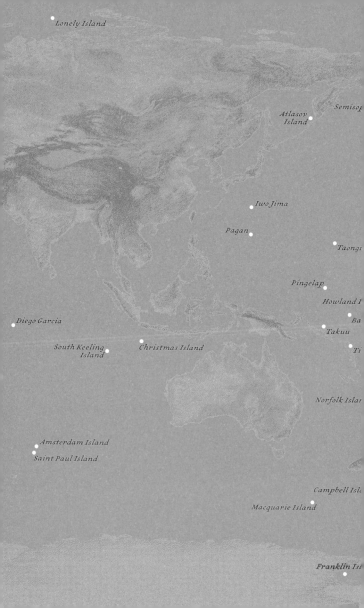